Administration of Desert and Drought-Prone Arid Areas

Changing Contours of Administration Since Independence
Golden Jubilee Volume Series - 13

Administration of Desert and Drought-Prone Arid Areas

Editors
Rakesh Hooja
R.K. Choubisa

Series Editor
Rakesh Hooja

The HCM Rajasthan State Institute of Public Administration, Jaipur

RAWAT PUBLICATIONS
Jaipur • New Delhi • Bangalore • Mumbai • Hyderabad • Guwahati

ISBN 81-316-0242-7

Published by
Prem Rawat for **Rawat Publications**
Satyam Apts, Sector 3, Jawahar Nagar, Jaipur 302 004 (India)
Phone: 0141 265 1748 / 7006 Fax: 0141 265 1748
E-mail: info@rawatbooks.com
Website: www.rawatbooks.com

New Delhi Office
4858/24, Ansari Road, Daryaganj, New Delhi 110 002
Phone: 011 2326 3290

Also at Bangalore, Mumbai, Hyderabad and Guwahati

Printed at Chaman Enterprises, New Delhi

Contents

Series Editor's Introduction

RAKESH HOOJA

It is most likely that all possible types of developmental, governmental, regulatory and administrative initiatives have been tried out in some part or the other of India over the last 60 years since the country gained independence. With governments of different ideological hues having remained in power at different points of time in different states of the Indian Union and at the Centre, and with different ideas about development, governance, planning, state welfarism, market-orientation, administration, nation-building etc. having been fashionable or accepted in different time frames, as well as a willingness to try out and emulate what was being attempted in countries of the West, or of the Soviet bloc, or in other developing areas, all forms of administrative and governmental action and all types of development schemes, as well as types of organisational arrangements, have been tried out somewhere or the other in India and have been commented upon by experts, scholars, researchers, practitioners, journalists and the ordinary laymen. In fact, Public Administration is a well-established discipline in India and has been regularly analysing changes in the field, as also in administrative thinking and in policy making, over the years. These debates, discussions, descriptions, analyses, comments, opinions, findings and administrative prescriptions have been duly covered in the various articles published in the two journals of the HCM Rajasthan State Institute of Public

Administration (HCM RIPA), which contain useful and valuable reference material.

It is in this context that we have undertaken this project of preparing thematic volumes, which would indicate some of the important changing contours of Indian Administration since independence. We wanted to look at how concepts, approaches, organisational structures, procedures, policies, programmes, projects, working styles, and modes of observation, analysis and evaluation, have all emerged and undergone change over time.

Accordingly, to celebrate 50 years of HCM RIPA, we felt it appropriate to select the best and most appropriate articles from the back issues of the over 34 years that our journal *Prashashnika* is being published, as well as the 11 years that our second journal *Development Policy and Administrative Review* used to be simultaneously printed. Deliberately wanting the articles selected for inclusion in this series to retain the flavour of the period in which they were written and to reflect the thinking and practice of that time, we have chosen not to modify the selected articles, except in some cases to slightly change the titles of the articles. The words, descriptions and analyses are of the authors as they had originally written the pieces. However, the articles selected have their intrinsic value, and by putting them together into separate thematic volumes, we have, hopefully, enabled readers to visualise how certain administrative concepts, or governmental and developmental programmes (or experiments) and policies, have been initially conceptualised and implemented, and have subsequently evolved over time. We have also attempted to provide a judicious blend of the theoretical and the practical in each book as also in the series as a whole.

Having gone through the back issues of *Prashasnika* and of *Development Policy and Administrative Review*, we have selected the following as the themes for each of the separate volumes:

1. Development Administration and Rural Development
2. Training, Civil Services and Personnel Administration
3. Panchayati Raj, Participation and Decentralisation
4. Governance and Reforms
5. Welfare and Tribal Development Administration

6. Administration of Public Sector Enterprises and Industrial Development: Changing Trends Since Independence.
7. Administrative Theories: Approaches, Concepts and Thinkers in Public Administration
8. Prashasnik Parivartan or Administrative Change (in Hindi)
9. Plan and Project Formulation and Implementation: Theoretical and Applied Aspects
10. Urban Development and Administration
11. Administration for the Empowerment and Welfare of Women
12. District Level Administration and the Role of the Collector
13. Administration of Desert and Drought-Prone Arid Areas
14. Administration of Water Management and Command Area Development
15. Administration of Health and Education Sectors
16. Managing Change, Organisational Behaviour and Human Resource Development
17. Land Use, Revenue Administration and Land Reforms
18. Criminal Justice and Police Administration
19. Financial Management and Economic Administration in Government

Other titles may be added as we go along.

I have received able support from my colleagues on the Series Editorial Board, namely Dr. Ramesh K. Arora, Shri S.N. Lathi and K.K. Parnami, in finalising the themes for the individual volumes and identifying the possible thematic volume editors. Mr. Parnami got a cumulative index prepared of all matter published in our two journals without which this exercise could not have been started. We are also gratified that the two publishers Mr. Subodh Bhushan Gupta of Aalekh Publishers and Mr. Pranit Rawat of Rawat Publications respectively, agreed to jointly co-publish the books with HCM RIPA at no cost to HCM RIPA, and to distribute them. The editors of each volume also are to be thanked for their efforts at meticulously culling out and selecting the best and most relevant articles for our consideration for inclusion in the respective volumes, and for their writing appropriate volume introductions.

It is hoped that this series of volumes shall prove to be a useful source of reference contributing to better understanding of the themes covered and acting as a stimulus for further reflection to all those interested in Indian administration, and that readers shall enjoy and appreciate reading the books as much as we in the Series Editorial Board, and the volume editors, did while preparing the books for publication.

Volume Editors' Introduction

RAKESH HOOJA AND R.K. CHOUBISA

Life in arid desert lands is a perpetual struggle where the consequences of proper resource management can be extremely grave. Each succeeding generation of desert dwellers, and of those who administer desert and drought-prone areas, has to adopt to life in the desert based on available technologies which have been found to be appropriate and adoptable for the prevailing conditions. While the modern era's scientific and technological breakthroughs have opened up new possibilities, any abandoning of traditional knowledge based and time tested practices and of the use of social organisations and social capital for marshalling available natural resources and for coping with life in such arid areas, is fraught with dangers. Natural resources of the desert regions of the country are under great stress on account of the increase in human and livestock population. The major problems include continuous depletion of vegetative cover, increase in soil erosion, fall in ground water level and increasing pressure on pastures and grazing lands for the livestock. All these factors account for diminishing productivity of the land and loss of natural resources. It is, therefore, necessary that the adverse effects of desertification and adverse climatic conditions on crops and on the human and livestock population are mitigated.

Policy and programmes for the development of desert areas must aim at restoration of the ecological balance by harnessing, conserving and developing in a sustainable manner the natural resources like land, water, vegetative cover, livestock and at raising land productivity in India. Conservation and optimal use of water and effective land water management are the prime requirements of life as well as governance, in desert, drought-prone and dry (rainfed farming) areas. The nature of agriculture and livestock bearing in such area is conditioned by this. Techniques of desert irrigation, water harvesting and watershed development, grassland management, afforestation, selection of appropriate crops; in fact the strategies of agriculture and rural development for desert areas – all have to be geared to coping with the desert environment. Even the health problems are somewhat different in the desert areas.

Integrated desert development approaches have to be worked out and implemented by a governmental delivery system specially oriented, structured and geared for this task; and one which encourages community participation. And droughts, which strike most western Rajasthan districts at least once in 3 or 4 years, need their own response mechanisms, governmental as well as community group based, to be put into operation.

An attempt has been made in our selection of articles for this volume to provide a cross section of problems being faced in this arid and semi-arid environment and to review of solutions evolved in the past as well as being attempted today.

Development of Desert Areas

While discussing the problems of desert, H.S. Mann has emphasised the need to adopt a scientific approach for arresting the desert. In his opening article he has rightly pointed out that the implementation of land use plans requires building up public opinion together with legislation and administrative measures. As agricultural practices in low rainfall arid areas can be injurious to the landscape, there is a need to develop high yielding ranges in desert areas. He suggests that tourism and wildlife should be more exploited to improve the general standards of living of the desert people. Based on his experiences at Central Arid Zone Research Institute, Jodhpur (CAZRI), Mann has vividly explained the strategies for sand dune stabilisation, afforestation, shelterbelt establishment, rangeland management, arid horticulture, crop production and rodent pest management. He is of the view that an integrated natural resources survey should be undertaken in the desert areas.

Discussing the social aspects of desert development, S.P. Malhotra examines traditional systems of resource use. He feels that the sources of livelihood and the way of life of the sedentary population had been conditioned by the disturbances in the ecological balance owing to the vagaries of the climate. He has examined in detail the problems of people living in desert areas like livelihood, working as casual labour; agriculture and minimising the risk of being affected by crop failures through subsidiary livestock rearing, scattered agricultural holdings, judicious crop rotations, moisture conservation etc.; social-religious ceremonies; and indebtedness. He asserts the view that we need to impose checks on the relatively high increasing population in the arid zone of Rajasthan. According to him, high fertility norms and social values pre-disposed for having more children need to be changed to reduce fertility. Therefore, he is in favour of developing avenues of employment other than agriculture. Malhotra has studied nomadic and semi-nomadic life of inhabitants and their relationships with the settled population. Based on his studies on the nomads, Malhotra gives certain recommendations, which are quite useful for improving their social life.

K.A. Shankarnarayan, in his article on monitoring of desertification, warns about the indiscriminate exploitation of nature by people. He examines the impact of exploitation in relation to desertification in terms of geomorphology, geo-hydrology, soils and soil fertility, shifting sand dunes, increased sand dunes, and degradation of vegetation changes in water balance. He found evidence of excessive exploitation of groundwater resources in some desert areas. Shankarnarayan has studied ecological changes in Luni development block of Jodhpur where he found the presence of the recognised ecosystems, their susceptibility towards deterioration, and extent of degradation by projecting species composition and biomass production. He is of the view that the desertification is very intensive; there is a need for diagnosing and monitoring desertification in the state. He has dealt with the process of desertification monitoring and approaches to undertake monitoring activities. He concludes with some ecological principles for combating desertification.

In his article on integrative machinery in the desert development administration in Rajasthan, Dr. Ramesh K. Arora examines the set up of administrative system in Rajasthan in which a Special Schemes Organisation (SSO) (*f.*1973) used to function for many years at the state secretariat. It used to be the most prominent agency for the planning, management and monitoring of desert development programme. Administrative structure and functioning of this multidisciplinary body has been explained in detail. Desert Development Commissioner with his headquarters in Jodhpur was another state level agency who supervised, coordinated and monitored the desert development programme in the desert districts. District rural development agencies at the district level implemented the programme. Arora has discussed some noticeable limitations in the realms of control, coordination and responsibility in the desert development administrative system. He favours integration in the administrative structure for desert development. Arora concludes with a number of suggestions and recommendations for redesigning and adaptation in the system.

Rakesh Hooja, in his article, discusses the developmental delivery system as it existed in desert districts in the 1980s. He has dealt in detail with the functioning of rural development administration at the district level. He covers constituents of the delivery system keeping in view various rural development programmes. These include revenue agency, departmental staff, development agency, new agriculture extension agency, special project agencies, institutional credit system, voluntary agency, beneficiaries associations, normal trade channels and the mass media network. Next, Rakesh Hooja discusses in detail the working of the delivery system in a desert district where rural development programmes are being implemented. He describes general problems of the rural development and desert development delivery system and their possible solutions.

Harsh Sethi has studied area development programming in western Rajasthan. He has examined development problems of a backward area with the help of the techniques of regional planning. He is of the view that the effectiveness of a regional development programme should be viewed not only in terms of scarce resources but also in terms of its induced effects on locally available idle resources. Sethi has looked at specific problems of the desert areas of western Rajasthan. To him, an intensive survey of the type that NCAER had undertaken should be conducted for the districts of Bikaner, Barmer, Jalore and Jaisalmer. An attempt should be made to describe how the present economy hangs together; and special emphasis should be placed on examining the potential of increase in dry farming, animal husbandry and allied industry, rural handicrafts and small-scale mining.

Next, R.P. Singh analyses alternatives to combat desertification in drylands. He warns that desertification leads to loss of productivity of the resource-base, deterioration of the health and well being of the people at risk, and desertification-induced migration and resettlement. He begins with managerial aspects of drylands ecosystems. Some measures to be planned include establishment of wind breaks and shelter belts; strip cropping of grasses and crops; stubble milch farming; water harvesting and its recycling; installation of small water and sediment-retaining dams, check dams in gullies; and planting of quick growing cover on devegetated areas and gullies. Early onset of monsoon, late arrival of monsoon, early recession of monsoon and breaks in monsoon are some of the aberrations that occurred in Rajasthan from 1901 to 1980. He studies these aberrations in detail and suggests that a study of weather aberrations on a long-term basis is essential for formulating a sound crop production strategy for drylands. He concludes with the notion that drylands ecosystems are highly fragile. Mismanaged or carelessly managed dryland ecosystems lead to desertification. He is of the view that following the proven soil and moisture conservation procedures, contingency plans should be followed according to varying weather aberration situations. He therefore suggests that emphasis should be laid on low cost monetary inputs or technologies, since the farmers inhabiting these drylands are poor.

L.D. Ahuja in his article, "Grassland Management for Desert Reclamation and Rangeland Development", examines the biophysical conditions of desert areas of Rajasthan by giving an analysis of livestock population in different agro-climate zones in arid districts of Rajasthan and its pressure on land. His findings are based on his studies on rangeland sub-stations under different agro-climate conditions conducted under the Central Arid Zone Research Institute, Jodhpur. He begins his analysis by listing the classes of rangelands in arid regions of the state. He discusses in detail several techniques of improving rangelands. These include: selection of sites, fencing, forage production, grabbing of unwanted bushes, top feed-cum-shade trees on grasslands, soil conservation, reseeding, fertilisation, utilisation, and livestock production and development.

Ahuja feels that this is not easy and therefore there are some constraints. To him, with limited rainfall in arid regions, the sowing period for agricultural crops and for reseeding of rangelands is very small and these coincide with each other. Besides, rangeland development work is mostly concentrated on land under class V and above, which have rough terrain, are subject to erosion hazards and therefore have disturbed vegetative cover. Their cost of development is higher than that of well developed agricultural lands. Ahuja has given a couple of suggestions to stop social and economic evils resulting from erosion hazards in the present desertic areas.

B.S. Chauhan has discussed public health problems in desert environment in the next paper. He is of the view that public health problems in desert environment are peculiar and different than in non-desert areas, due to the nature of deserts. Climate exerts a great influence on the health and socio-economic development of man. He suggests *desert medicines* for different diseases in desert areas. Chauhan feels that diseases in the deserts are mostly due to geogens rather than pathogens. He gives a list of diseases, which are common in a desert environment. He calls for a health programme that provides a community diagnosis and treatment and economises cost of health services in sparsely and scattered populated areas and for making it "health for the people, by the people and of the people". He suggests that health planners for desert regions should stay in the area and work for the people. Their work would then approximate more to the aspirations of the people and they will be able to involve the community in solving the problems.

Water and Watershed Development

Water is our earth's most precious natural resource. It is a basic human need and a precious national asset. Water is a "cradle of life" on which all organisms play. It is one of the basic needs of life and essential for survival. In recent years there has been a growing perception of a looming water scarcity, which is particularly acute in desert areas.

Rain-fed areas constitute about 60 per cent of net sown area in the country and are characterised by low levels of productivity and low input use. The bulk of India's rural poor lives in rain-fed regions and face high variability of rainfall, resulting in wide variation and instability in yields. For sustainable development of these areas, the watershed development approach has been adopted and given high priority for several years.

As C.H. Hanumantha Rao had stated in his well-known Loveraj Memorial Lecture of 2000, "watershed development has been conceived basically as a strategy for protecting the livelihoods of people inhabiting the fragile ecosystems experiencing soil erosion and moisture stress. The aim has been to ensure availability of drinking water, fuel wood and fodder and to raise the income and employment of farmers and landless labourers through improvement in agriculture production and productivity."

Similarly Ministry of Rural Development (Government of India) Guidelines for Watershed Development 2001 points out that "the watershed approach has conventionally aimed at treating degraded lands with the help of low cost and locally accessed technologies such as in situ soil and moisture conservation, afforestation etc., and through a participatory approach that seeks to secure close involvement of user committees."

The watershed approach has been a major programme of drought-prone areas' development and desert development efforts in the country. Rajasthan also relied on watershed development and the massive Indira Gandhi Canal (formerly Rajasthan Canal) for bringing water to Rajasthan's sand dunes and making desert irrigation possible.

Kanwar Sain, the renowned irrigation engineer, in his paper considers water as the most effective tool in the development of desert areas. He traces the history of Gang Canal, which ultimately led Rajasthan state to have the Rajasthan Canal Project (later renamed Indira Gandhi Canal Project). Sain discusses briefly the development of Thar Desert area (now in Pakistan) since the inception of canal irrigation in 1870. Coming to the Rajasthan Canal, he discusses how it was initially conceived and initiated in the end of the decade of the 1950s. He also touches on the culturable area for irrigation and sources of water for the Rajasthan Canal. According to him, weather modification techniques can be applied to secure some additional water. He is in favour of a systematic research on the Rajasthan desert to determine the most effective method for inducing more rain in the cloud formative conditions in this area by scientifically designed and vigorously checked experiments and to evaluate the amount of precipitation resulting from such experiments. He concludes with a summary of schemes envisaged by the Central Water and Power Commission, which on the one hand led to the commencement of the Rajasthan Canal Project, and on the other proved to be the precursor of the interlinking of rivers project being considered seriously in the opening years of the 21st century.

Anirudh Krishna has outlined a scheme for dryland areas in Rajasthan. Since 96.5 per cent of Rajasthan's land area is dependent on rainfall for all its requirements of moisture, the conservation and harvesting of rainwater is of great significance in the state. Watershed-based development is the option he stresses upon for rainfed areas. The National Watershed Development Projects for Rainfed Areas (NWDPRA), launched during the VIII Plan, brought a new Watershed Development and Soil Conservation Department into the existence in Rajasthan in January 1991. Krishna discusses the background in which the new department was established. He also describes the methodology of work that the new department adopted. Selection of watershed, watershed treatments, project objectives, people's participation, users' committees, Mitra Kisan and Gopal, participating planning, emphasis on training, learning from other states, topographical surveys, concurrent evaluation and research supports are the components of the NWDPRA. In the end, Krishna discusses several administrative issues that the newly created department faced in its initial stages.

Rakesh Hooja and Rakshat Hooja have highlighted the role of community-based organisations and panchayats as instruments of governance in the sphere of watershed development. They have summarised the principal features of institutional and administrative arrangements for watershed development as contained in Common Guidelines of 2001. Ministry of Rural Development, Government of India has also launched a new *Haryali* scheme in January 2003 under which it replaced the Common Guidelines with the *Haryali* Guidelines. Authors have also indicated some changes brought by this new scheme. Community-Based Organisations (CBOs) and Gram Panchayats are the instruments for watershed development. They feel that greater emphasis upon CBOs would enable the achievement of natural resource related land water management goals of watershed development. Appropriate mechanisms and procedures for promoting harmonious relations between community-based organisations and Gram Panchayats need to be innovated with and put into practice.

In his article entitled "Watershed Plus in Rajasthan", M.K. Khanna describes various components of watershed development and examines the situation of agriculture, animal husbandry, farming and water management in the state by giving a brief history of watershed development programmes started from 1995-96. He outlines broad objectives and key concerns of watershed programmes and gives valuable suggestions for improvement. He also provides a brief summary of Gandhi Gram Yojna, which is a unique village-based watershed and community development project.

Khanna also sets forth the basic approach of Watershed Plus philosophy of the Department of Land Resources of the Ministry of Rural Development. The new paradigm recognises the need to involve the community as a necessary condition for the sustainability of a watershed programme. He details some key action areas and an action plan for DPIP Plus in Rajasthan with some expected outcomes.

Rakshat Hooja examines the issue of water harvesting projects and the efforts made by Tarun Bharat Sangh (TBS), an NGO working towards community mobilisation for sustainable natural resources management in the state. He consolidates the experiences of the TBS in using popular/traditional/people's knowledge and participatory planning for water harvesting. He also discusses various stages, components and tools of the TBS methods. He indicates that there are four water holding capacity structures constructed by TBS in the villages: *Anicuts, Paal, Johads and Medhbandhi. Tanka, Khet, Talai* and *Bandha* are other structures in the villages. The study relates to villages structures created by TBS in Alwar, Banswara, Sawai Madhopr, Karoli and Jaipur districts. TBS has adopted a professional approach according to which they first create awareness. The Gram Sabha meets and selects a site at the next stage. A suitable structure is planned and households contribute their share either as labour or cash. TBS looks after the maintenance of structure. This type of experiment provides us learning and lessons.

Agricultural Development in Arid Areas

In the opening article of this part, Bhupendra Hooja presents a bird's eye-view of the up-to-date developments of agrarian structure in Rajasthan. Giving a historical background of the structure, B.

Hooja says that agriculture and land administration in the country were not imbued with the ethos of development. There had been evolved during the later period of the colonial rule, chiefly to augment the coffers of the empire, or to link up Indian agriculture to the world-wide but empire-dominated mercantile economy, and incidentally or partly to meet the recurring challenges of droughts and famines. According to him, the progress in agrarian system has not been satisfactory because of adverse climate conditions, frequent and repeated droughts, erratic or uncertain rainfall and occasionally heavy downpours and sudden floods. Yet, overall achievements have been significant. However, climate, irrigation potential and degraded lands have limited the growth, which have brought forth consequent challenges in the agrarian sector. Green and White Revolution, land and water management, Brown/Yellow Revolution, water discipline, animal husbandry and fodder development and a series of technological measures have had a tremendous impact on agriculture. Application of new technologies and building up new institutions has led agriculture towards modernisation. Bhupendra Hooja discusses the following new trends in agrarian sector: management of common lands, watershed management, autonomous units in water and IRDP. In the end, he proposes a revised organisational structure for agriculture.

In his article, Nathu Ram Mirdha has conceptualised area development authorities. He explains area development programmes like command areas of irrigation project, dry farming areas, drought-prone areas, desert areas, hill areas and tribal areas. According to him, desert areas are the most backward in our country. Doubtful rainfall, sand storms, sub-soil salinity and alkalinity in places and nomadic livelihood based on cattle and sheep have made these areas difficult for development. He advocates for designing area development programmes in such a manner that the programmes per se reinforce and supplement each other in the overall economic development of the area.

Bhupendra Hooja in his second article explores planning priorities and development policies in the agriculture sector. Starting with a philosophy of development in India, B. Hooja goes on to discuss the pioneering work of the National Planning Committee which set out some broad objectives for improvements in food, clothing and housing standards, increased production in agriculture and industrialisation, employment opportunities, increase in per capita income, extension of public utility services and educational as well as medical services. He further defines the role of the state in industrialisation. On the eve of independence, India saw Bengal famine, country's partition, Jammu & Kashmir and Indo-Pak conflicts. He gives a detailed picture of agricultural development in both pre-planning and post-planning era. He sets priorities in planning for agriculture development. These are: land development and scientific management, social justice, land reforms, cooperatives, dry-land farming and watershed projects, beneficiaries participation, integrated approach, and area development. Bhupendra Hooja is in favour of areawise budget on a local area to local area, or region to region, basis.

In his paper, S.P. Vishnoi has identified some problems of agricultural marketing in the desert areas of Rajasthan. Production, assembly, grading, storage and transportation, actual marketing of produce etc., are some of the problems in arid zones. According to him, overall poverty and low economic activity in arid and semi arid areas is the foremost problem. Total agricultural production in these areas is normally insignificant. Middlemen control the animal products and primary raw materials and producers are traditionally indebted. Adverse agro-climate conditions affect activities and viability of cooperative societies. Banking facilities are not satisfactory in arid and semi-arid areas. Producers do not enjoy the benefit of cheap and easy credit and institutional purchases are small. Vishnoi finds that the problems of the desert present different situations both at the planning and at the execution level. He suggests that the integration of various development programmes and schemes in this manner with more specific attention to the infrastructural constraints, seem to be the solution.

A.N. Lahiri discusses in detail the optimisation of crop production in desert lands. He identifies problem of wind erosion in arid areas. According to him, hazards of wind erosion ask for appropriate soil conservation measures, low and erratic precipitation warrants moisture conservation in sandy soils, poor and fertility conditions require methods for its maintenance and improvement, and meagre irrigation waters need effective management for optimum crop production with minimum use of water. Based on his study conducted in Beechwal and Udairamsar areas of Bikaner district, he takes into account consideration of soil moisture situations and plant water use in dry regions. While discussing management conditions for grasses, he finds that management may be tailored to meet the demands of high forage production. He is of the view that in view of the high density of human population in the Indian desert, the practice of cropping cannot be entirely eliminated. He contends that agricultural production in rainfed arid and semi-arid areas is highly unstable mostly because of frequent droughts. He suggests that several technologies like sub-surface barrier, use of mulches, runoff farming etc., should be developed for soil moisture conservation and for the avoidance of various kinds of droughts. He concludes with the view that the diverse technologies provide ample confidence for optimisation of production from dry regions. However, the crux of the problem is the effective diffusion of these technologies in rural areas. According to him rational use in reckoning with the land capability is another dimension of the problem, which warrants special reckoning in desert areas for the conservation of the fragile ecosystem.

C.M. Mathur comments upon forestry in western Rajasthan. The natural vegetation of western Rajasthan (Shri Ganganagar, Bikaner, Barmer, Jaisalmer, Jodhpur, Churu, Naguar, Jalore, Jhunjhunu, Sikar and Pali) is quite sparse with limited number of species. The forest forms a negligible proportion of the total geographical area and is even less than 2 per cent. The climate is characterised by high solar radiation, extreme variations of diurnal and seasonal temperature and erratic rainfall. The vegetation of this region is scrophytic scrub type. Therefore various schemes have been implemented under the desert development programme and as part of Rajasthan Canal Project (RCP). Mathur gives in detail forestry schemes like farm forestry, silvi-pastoral plantation, sand-dune stabilisation, shelterbelt-cum-roadside plantation and development of fodder bank. He gives physical achievements and financial outlay under desert development programme in initial two years. Then, he takes into account the afforestation programme in Rajasthan Canal Project area, which envisages protection of canals, roads and farmlands from shifting sand dunes and ensuring supply of fuel, wood, timber and fodder with the assistance of a World Bank project. He lists out various schemes like shelterbelt plantations along canals, shelterbelt plantations along roads, fuel wood plantations near villages, sand dunes stabilisation cum pasture development in uncommanded areas of project area and farm forestry programme under RCP through farmers. To Mathur, forestry operation in the desert region outside RCP area is quite a challenging task because there are many odds in the way. The magnitude of work taken up is very meagre compared to the vast fields lying barren in this area. He further contends that people's participation is very vital for all these works otherwise no programme will be successful at all.

Famines and Droughts

In India, an estimated area of 146.82 million hectares suffers from various kinds of land degradation due to water and wind erosion and other complex problems like alkalinity/salinity, waterlogging, soil acidity and so on. Famines, droughts and floods are also a common feature in many parts of the country.

Broadly speaking, drought-prone areas which cover both arid and semi-arid areas—where arid areas are districts with rainfall up to 375 mm [15 inches] and semi-arid areas are districts with rainfall from 375 mm to 750 mm [15-30 inches, an infect rainfed drylands having rainfall of less than 800 mm are almost all in arid or semi-arid areas]. In terms of geographical area and population, drought-prone areas account for nearly 19 per cent of the total area of the country and 12 per cent of the population. In terms of net area shown in India, 68 per cent is drought-prone. In some states such as Rajasthan, about 56 per cent of the total area and 33 per cent of the population fall in this category, Next to Rajasthan comes Andhra Pradesh, where about 30 per cent of the geographical area and 22 per cent of the population lies in drought-prone areas. In Gujarat about 29 per cent of the geographical area and about 18 per cent of the population are situated in drought-prone areas. In Karnataka, the corresponding figures are 25 per cent and 22 per cent. Excepting Kerala, Punjab and the states of north-eastern region, every state in India has one or more drought-prone areas.

In Rajasthan, there have been 48 severe drought years of varied intensity since 1901. A more detailed analysis reveals that only 9 out of 106 years were such when none of the districts in the state were affected by droughts. Between 2000 and 2003, Rajasthan experienced two drought years, which had severe effects on the social and economic lives of the people of the state. Accordingly, practically in every year some part(s) of Rajasthan is affected by drought.

A drought affects irrigation and crop production, food availability, fodder, calory intake, drinking water, livelihoods, incomes (even incomes in non-farm sector in both rural and urban areas since the rural and urban economies are interdependent), entitlements, rural indebtedness, health and malnutrition, life expectancies (of both human beings and animals), livestock, agro-based and allied activities, the state's and district's administrative machinery, and the overall economy of the region that faces drought. Drought may even cause several distress nutrition.

We feel that drought policy should be divided into four major components:

1. Drought proofing/drought prevention – this could also be referred to as integrated development and resource management in drought-prone areas;
2. Mitigation of impacts on droughts in years of moderate drought;
3. Drought relief operations in a crisis or campaign mode in years of severe drought;
4. Administrative structures, procedures and arrangements to deal with as also to manage and monitor (i), (ii) and (ii) above.

Turning to the articles in this section, Sanjay Sharma in his interesting paper discusses prediction of drought in arid Rajasthan folklore. To begin with, he defines terms like drought and famine. To him, drought means a considerably lower rainfall from normal. It is a situation occurring in any one area when the rainfall is less than 75 per cent of the normal. When deficiency of rainfall is about 50 per cent of the normal it is termed as a severe drought. On the other hand, famine is a state of affairs in which people of a region suffer from acute shortage of food, fodder and water or any one of them. Famines are thus classified as *Annakal* (grain famine), *Jalakal* (scarcity of water), *Trinakal* (scarcity of fodder) and *Trikal* (scarcity of grain, water and fodder). Sharma has tried to suggest how the people of the arid region have over the years evolved and passed on the indigenous knowledge about predicting the onset of monsoon, drought and famine, obviously a response to meet human needs. By using a number of sayings and verses in Marwari he tries to narrate the predictability of monsoon in arid Rajasthan.

Kamala Jain deals with the famine code and its administration in the 19th century. She starts with famine relief policy of Rajputana during the later half of the 19th century. It was through its resolution of 24 September 1868 that the British government in India laid down the policy, to be followed by the princely states, to meet the famine crisis. Shortly thereafter a Famine Commission was appointed by Lord Starchy, which was asked to examine the famine conditions in princely states. The Commission eventually drafted a provisional Famine Code in 1883 and later a Draft Famine Code for princely states in 1885. She describes the main contents of the Draft Famine Code. The provisions of the Code were duly implemented in Rajputana. Public, irrigation and communication works were undertaken. The famine of 1899-1900 known as *Chappanya Kal* that occurred in *Vikram Samvat* 1956, was a Trikal in Rajputana. Jain has done a comparative study of relief operations during the famines of 1868-69 to 1899-1900 in Rajputana. The study shows considerable improvement—both in the conception of the ruling chiefs as regards their responsibilities towards the people—and in the measures adopted by them. In the end, she recommends that a correct and long term famine policy should aim at banishing poverty and unemployment among the masses so that the purchasing power of the poor masses is increased.

J.P.Kanawa gives a blueprint of famine, drought and scarcities in Rajasthan. He starts with defining famine zones and districts of the state and goes on to indicate causes of famine and droughts. To him, the causes of famine are basically natural but economic, social and political factors are also responsible for the existence of this problem. He analyses government measures to solve the problem. He states the DPAP and DDP are short-term strategies for drought management and that extension of irrigation facilities, drainage facilities, effective coordination between various programmes, stabilisation of sand dunes, and promotion of non-farming activities are long term strategies for drought management. He concludes that there should be proper coordination between short-term and long-terms measures to solve the problem of drought and famines in future; that the cooperation of NGOs and voluntary agencies would go a long way to mitigate the problem of droughts and famines in the state; and that public cooperation is a necessary condition to fight famines and droughts successfully in the years to come.

I.C. Srivastava starts with a brief historical background of famines and droughts before and after independence. He considers the present status of agriculture and the avenues of employment in other sectors. The salient features of Indian economy have been narrated. Srivastava thereafter discusses in some detail various issues like famine relief works, creation of employment, redistribution of income to the poor, generation of sustainable livelihood opportunities for the poor, avenues of employment, promotion of agri-business activities, absorption of the rural poor in other economic activities, rural credit, promotion of self-help groups, rural transformation, internet-based real-time information, customised knowledge, ITC plans, e-governance services, development of rural entrepreneurship, area-specific action plan, capacity building, and training of personnel. Srivastava feels that with the above-mentioned efforts we may face famine fearlessly. Improvement of service delivery approach is the way through which we may combat recurring and frequent famines.

References

Farrington, J., C. Turton and A.J. James (ed.) 1999, *Participatory Watershed Development*, OUP, Delhi.

Hanumantha Rao, C.H., *Watershed Development in India: Present Experience and Emerging Issues*, Loveraj Kumar Memorial Lecture, Society for Promotion of Watershed Development, New Delhi.

Hooja, Rakesh, 2006 (a), "Drought Policy Framework" in *Indian Journal of Public Administration*, Vol.52, No. 1, January-March 2006.

Hooja, Rakesh, 2006 (b), *Management of Water for Agriculture: Irrigation, Watersheds and Drainage*, Rawat, Jaipur.

Hooja, Rakesh and R. Joshi (ed.) 1999, *Desert, Drought and Development: Studies in Resource Management and Sustainability*, Rawat, Jaipur.

Hooja Rakesh *et al.* 2006, *Drought Management and Administrative Reforms: An All India Perspective*, Centre for Management Studies, HCM Rajasthan State Institute of Public Administration, Jaipur.

Indian Council of Social Science Research, 19977, *Desertification and its Control*, New Delhi.

Jaiswal, N.K. and N.V. Kolte, 1981, *Development of Drought-prone Areas*, National Institute of Rural Development, Hyderabad.

Mann, H.S. (ed.) *Arid Zone Research and Development* 1980, Scientific Publishers, Jodhpur.

Mensching, H.G. and R.C. Sharma (ed.) 1984, *Resource Management in Drylands (The Rajasthan Example)*, Rajesh Publications, New Delhi.

MYRADA and IIRR, 1997, *Resource Management in Rainfed Drylands: An Information Kit*, MYRADA, Bangalore.

National Commission of Agriculture, 1974, *Interim Report on Desert Development Programme*, Government of India, New Delhi.

OIKOS and IIRR, 2000, *Social and Institutional Issues in Watershed Management in India*, OIKOS, India

Spooner, B. and H.S. Mann, 1982, *Desertification and Its Control: Dryland Ecology in Social Perspective.* Academic Press, New York.

I

Development of Desert Areas

1

Scientific Approach for Arresting the Desert

H.S. MANN

The association of actions for combating desertification and land use patterns in arid lands is based on a fragile balance which can be further upset. It is essential, therefore, that plans of development should be based on maintaining optimal sustainable productivity, application of ecological and socioculturally sound technologies that will not upset the ecological balance, and development of innovative modalities for administration and extension services.

The implementation of land use plans requires building up public opinion together with legislation and administrative measures. Strong demographic policy implementation is warranted without further loss of time regarding control of human as well as livestock populations since man is becoming increasingly dependent upon arid and semi-arid lands.

Agricultural practices in low rainfall arid areas are injurious to the landscape and aggravate the sand movement. Considering this evaluation, stress should be given to development of high yielding ranges and to professions which do not require the use of water.

Surface water development achieved by the desert dwellers through their ingenuity and long-standing efforts need to be further strengthened by development of modern technology and resources. There is a large scope for improvements in water management in already existing and on-going projects.

Not only the processes of desertification should be monitored for environmental impact assessment but also the implications of implementation of new technologies.

The technologies available at present are not reaching the rural farmers/industrialists as fast as they should. Much more attention is to be paid by the desert states on this aspect.

Contingent plans for drought periods and for excessive rainfall years should be thoroughly planned. We should not be taken unawares.

The distance between desert scientists (who develop technologies), administrators and politicians (who formulate the policies and take decisions) should be shortened.

The employment programmes instituted during drought relief should have far-reaching implications to augment the productivity of the arid lands involving the rural masses.

Tourism and wild-life should be more exploited to improve the general standards of living of the desert people.

The Strategies

Besides creating a highly scientific environment, a large number of successful technologies have been evolved at CAZRI and demonstrated at about 60 centres spread over entire western Rajasthan, keeping in view the above mentioned approaches. Some of them are explained here.

Sand Dune Stabilisation

Silvi-pastoral technologies developed and demonstrated at CAZRI for stabilising the bare sand dunes which pose as a menace to human habitation and agricultural fields in the desert areas, consist of:

(a) protection of shifting dunes from biotic interference.

(b) establishment of micro-wind breaks on the wind-ward side of dune in 5 m parallel strips or 5 m^2 chess boards, and

(c) sowing of grasses and transplanting (with the onset of monsoon) of adapted trees and shrubs raised in earthen bricks on the leeward side of micro-wind break.

Economic analysis of this stabilisation has indicated that the average cost of Rs.760 per ha will be repaid after the end of thirteenth year. It has been found that about 50 to 70 mm of moisture initially present within a metre depth of unstabilised dunes helps in the seeding establishment. Moisure received in monsoon is adequate for seedling established for subsequent growth. Soil analysis after few years of establishment indicated that the technique influences the organic matter build in the sand dune.

Afforestation

Afforestation can be planned for several purposes. If the requirement is only for establishing green belts in the arid zones even the indigenous trees like *prosopis cineraria, azadirachta indica, albizzia lebbek, salvadora persica, oleoides, acacia nilotica, Arabica, a.n. leucophloia, a. jacquemontii*, which are of proven adaptability, though relatively slow growing, can be selected. However, if fast turnover is planned as for energy plantation, quick growing exotic trees like *eucalyptus, acacia tortilis, propspis juliflora* should be planted. A number of practical methods have been developed on the basis of (i) selection of species for different sites, (ii) seedling production, (iii) sil-vical characteristics of selected trees and shrubs, (iv) soil working techniques and cultural operations, and (v) protection against diseases and vertebrate pests.

In pursuance of the above objectives, 112 species of *eucalyptus*, 65 species of *acacia*, eight species of *cassia* and 63 species of 44 other genera were introduced in Jodhpur (mean annual rainfall 366 mm) and Pali (mean annual rainfall 411 mm) from Israel, Australia, the USA, the USSR, Southern Rhodesia, Chile, Peru, Sudan, etc. These studies have shown that among *eucalyptus, e. melanophloia* has greatest promise in Jodhpur, whereas in Pali, four species, namely, *e. tessellaris, e. menaophloia, e. collabah* and *e. camaldulensis* have shown good performance. Among the acacias studied, *a. tortillas* (Israel), *a. sieberiana* (S. Rhodesia), *a. seyal* (Kenya), *a. aneura* (Australia), *salicina* (Australia), *a. ligulata* (Australia), *a. victoriae* (Australia) have good performance. Of these, *a. tortilis* has proved to be the most promising (100% survival at Jodhpur and Pali) and did extremely well in shifting sand dunes.

In view of these results, *a. tortilis* has been introduced extensively in north-west India. Its performance has been equally good in Ramnad, Tamil Nadu and black soils of Maharashtra.

Among species of other genera, *colophosperum mopane* (S. Rhodesia) and *dichrostachys glomerata* (Israel) are doing very well in the Indian arid zone. One important characteristic of these two species is their self propagation. *Dichrostachys* naturally spreads through root suckers and *colophosperum mopane* by seed dispersal. Care should also be taken while selecting plants about their suitability to local ecological conditions and soil profile.

Planting-pits of 60 × 60 × 60 cm should be dug out and half filled with the loose weathered soil before planting with the onset of rains in the month of July. Sturdy seedlings of nine to ten months old, raised in galvanised iron containers or polythene bags, should be planted in the pits provided with saucer like depressions around the plants in plains or crescent shaped ridges of 15 cm high are formed across the local slope if planting is to be done on slopes so as to arrest the run off water during monsoons. Afforestation of saline areas is only possible by ripping open the pan to a depth of a metre by a tractor (D.C.8 crawler type) followed by cross harrowing. After the pan is broken, the area should be cross harrowed to break the crust formation. After this operation, ridges of one metre wide at the base and one metre high are constructed with the help of dosers. The inter-ridges space is about 1.5 to 2.0 m. Planting is carried out on the descrest of ridges in pits so that the salts get leached down from the ridges to lower depth providing better soil conditions and rooting depth for increased transplant establishment and growth.

For transplanting trees, during the short-lived monsoon, it is necessary that a very large number of saplings are raised in nurseries. Studies on the media and sizes of receptacles for raising saplings indicated that the use of metallic containers not only results into water economy but also enhances survival percentage compared with baked earthen pots. The use of metallic containers effected an overall economy of Rs. 6 per 100 seedlings. The metallic containers have now been replaced by cheaper polythene tubes. For planting trees or shrubs, the seedlings should be raised in specially prepared sun dried bricks, consisting of mixture of sand, clay and farm yard manure in equal proportion with a height of 30 cms and a cross section at top and bottom of 10 and 15 cm^2

respectively. A high degree of establishment is achieved if saplings are planted in this manner.

With the increase of human population, more and more marginal lands are coming under the plough though it may not be economical. A recent evaluation considered large scale tree plantations as an alternative to annual cropping. Five tree species *acacia tortilis, albizzia lebbeck, prosopis cineraria, prosopis juliflora, zizyphus* (grafted) were identified and the economics of raising each of them over one felling cycle was worked out. Discount factors at 11 per cent were used to compare the net annual returns from the tree species with those from the annual crop. Depending upon the species and the felling cycle chosen, the net annual returns from the tree crops were found to range from Rs. 360 to Rs. 3,270 per hectare. This was found to be much better than the expectation from the annual crops. Besides the higher financial returns from the tree crops, a number of social benefits like fuel, enhancement of wild-life and out-door recreation are also expected to emerge out.

Shelterbelt Establishment

Vegetative barrier placed in the path of the wind reduces its velocity near the ground by exerting a drag on the wind and deflecting the wind stream. This action constitutes the mechanism of possible wind erosion control. The effectiveness of shelterbelts in reducing wind velocity depends on many factors such as wind velocity itself, direction, and shape, width, tree height and tree density. The shelterbelts effectively protect crops and control sand drifting, to furnish cover and food for the wild-life etc. They are planted across and on the margins of agricultural fields. Successful shelterbelts *acacia nilotica* and *dalbergia sissoo* over a length of 102 kms were established at the Central Mechanised Farm, Suratgarh in Rajasthan. The soil working consists of digging ditches of triangular cross-section, i.e., 45 cm wide and 45 cm depth along the planting line with a mechanical ditcher. Planting pits of $30m^2$ are dug staggered with each other on either side of the ditches. Seeding of *a. nilotica sp. Indica* was carried out on either side of the ditch such that the sown lines remain just above the planted in the pits. The cost of raising such shelterbelts worked out to Rs. 330 per km.

Experimental shelterbelts as roadside avenues along the highways were established in western Rajasthan covering over 200 kms at a cost of Rs. 625 per km.

The technique consisted of planting three staggered rows of trees on either side of the road. The preface of the species in the order of survival is as follows: *prosopis juliflora, azadirachta indica* and *albizzia lebbek.* Where rainfall was 35 cm and above the survival of these species was over 95 per cent.

Rangeland Management

Considering that livestock husbandry occupies the most important place in the economy of arid region, the rangelands have an important role to play in the sustenance of livestock. Because of high density of livestock and therefore the grazing pressure, the grassland have deteriorated which not only cause serious imbalance in the supply-demand position of forages but also create severe soil erosion problems. Studies on rangeland improvement, management and utilisation have revealed that by simple closure of an area, the forage yield from the rangeland could be doubled in about three years time.

Soil and moisture conservation measures: On shallow soil with rolling topography laying contour furrows of 929 sq. cm. cross section (61 cm wide and 23 cm deep) at a distance of 8-10m across the slope increased the forage yield by 636 per cent. Contour furrows are cheapest to maintain and cost about Rs. 150 to Rs. 200 per hectare. For water harvesting in pasture land, a tractor drawn pitting disker has been designed and fabricated. Use of the pitting disker in cultivated pasture of *c. cillaris, c. setigerus* and *lasiurus sindicus* gave an average increase of over 30 per cent forage yield over control, and in the case of natural grasslands the forage yield increase was over 160 per cent. The fabrication cost of disker comes to below Rs. 400 if old discarded discs are used and Rs. 950 if new discs are used. The managed pasture should be protected against indiscriminate grazing through barbed wire fencing.

Reseeding: Owing to erratic rainfall conditions, and poor germination of seeds of *lasiurus sindicus* and *dichanthium annulatum,* the sowing of the mixture of seeds of *cenchrus* species and *lasiarus sindicus* and of *cenchrus* species and *dichanthium annulatum* is recommended for low and high-rainfall tracts respectively. The seed-rate of different mixtures is 5 to 7 kg per hectare. The requisite quantity of unhusked seeds of grasses, suiting the tract, should be mixed with moist sandy soil three to four times the volume of seeds and drilled uniformly in lines, 75 cm apart, immediately after the first effective showers in 8 to 10 cm deep furrows at a depth of not more than one to three cm under the soil. Such reseeded rangelands need at least two weedings during the first year of their establishment.

Introduction of legumes: As the availability of fertilizers is a problems and their cost is quite high, the role of legumes on grassland production is under study. In limited trials, legumes, in combination with cultivated pastures, have resulted in a higher nutrient (specially crude protein) content in the composite forage material. Studies on the compatability and performance of promising legumes, e.g., *dolichos lablab, phaseolus atropurpureus, cyamopsis tetragonoloba, phaseoulus aurea, p. aconitifolius* and *vigna sinensia,* on grasslands under different agro-climatic sub-regions are in progress. In areas, e.g., the Bikaner district, in pure stands of *losiusus sindicus*, legumes, e.g., *phaseolus aconitifolius* (moth) is taken as an interspace crop. It improves production and adds to the resultant product.

Silvi-pastoral approach: Considering that livestock husbandry occupies the most important place in the economy of the arid region and that frequent droughts result in a loss of cattle wealth owing to the shortage of fodder resources, it is necessary that range improvement should be complemented with the raising of fodder tree and shrub species which not only give the much needed forage during the scarcity periods, but also provide shade and shelter to the grazing animals, thereby helping them to utilise forage uniformly on the range. In addition, fodder trees and shrubs will ameliorate the micro-climatic conditions conducive to the natural regeneration of grasses which are higher in succession. Studies have shown that a density of 14 per cent of *zizyphus nummularia* is optimum for increased forage production. *Prosopis cineraria* is an important forage-tree species which grows in cultivated fields (sometimes more that 60-80 trees per hectare) without any detriment to the crops grown in association with it. The tree is lopped for its protein rich (17.49 per cent) leaves. Studies have shown that complete lopping (lopping of the entire tree) gives a significantly higher fodder yield (58-72 kg per tree) than the lopping of the lower two-thirds (28-48 kg per tree); and the lower one-third (19.73 kg per tree) of the crown.

Arid Horticulture

Suitable cultivers of *ber*, date palm and pomegranate have been identified. The selected *ber* cultivers yield as 45-50 kg fruit per tree even under the conditions of precarious rainfall, marginal soils and nominal inputs. Even in the absence of irrigation resource, the desert farmer can now have viable source of income and nutrition from *ber* cultivation. Besides, a run off concentration technique has been developed to enable the farmer to have an assured and satisfactory production even in drought years. A major bottleneck has been overcome making a breakthrough for quick raising of *ber* orchards by the development of a propagation technique. This technique takes not more than five months in raising budded plants which also establish with outstanding success even in xeric conditions. This technique with emphasis on rotation of the tap root and its straight and deep growing tendency can also be used for raising other fruit plants for the arid agro-climate. The technique has been used for distribution of *ber* plants to various organisations/ institutions in the country and for the Operational Research Programme of the CAZRI and has now been adopted by private nurserymen and extension agencies.

Crop Production

Limited water for irrigation: The strategies of optimum use of limited irrigation could be developed by linking low level of irrigation, with optimisation of inputs. Spreading of water over a large area appeared to be advantageous as the total yields are more from the whole area despite a lower yield per unit of area. Introduction of trickle irrigation system met the broader objectives of limited and gainful use of water, even where the water is brackish and its advantages over sprinkler and other methods of irrigation could be unequalvocally provided for a number of crops like potato, watermelon, ridge gourd, round gourd, tomato, etc.

Strategies for aberrent weather: In the event of pre-monsoon showers or early onset of monsoon in June, crops like *bajra* (BJ 104), *mung* (S-8) and *til* (T-13) could be planted with advantage. In case of delayed onset of monsoon (in the second half of July), *til* (T-13) is preferable over other crops. Low incidence of phyllody was recorded in *til* sown late in the season. In the event of failure of normal sown crop due to prolonged drought conditions and revival of monsoon in the first half of August, planting of *moth* and *guar* beans is advantageous over other crops like *bajra* and *mung*.

Crops and varieties: S-8 variety of *mung* has been identified as promising. It takes just 65 to 70 days to mature as compared to 84 days maturity period of RS 4 and has deeper root system, thus matching the soil-climatic conditions of this region. The variety has been adopted by farmers for cultivation over large areas in the region. *Guar* and castor, the latter being a new introduction have proved to be promising for stabilising production in low rainfall/ drought years. These crops have deep root systems and, hence, can make use of the moisture stored in the sub-strata, i.e., murrum layer.

Use of surface mulches: Bajra husk, a waste material, when applied as surface mulch at the rate of 4-5 t/ha as 80 to 100 per cent cover soon after sowing, has shown promise for use as surface mulch. The hazards of surface crusting and high soil temperatures are also reduced.

Use Potentiality of Poor Quality Groundwater

The groundwater in the region are generally saline to highly saline. Their use in irrigation does pose some problems. Since, invariably these are the only sources of water available, it is of immense practical interest to know what crops and their yields can be expected and with what managements. Work done at the CAZRI over the past several years has been able to establish this use potentiality of the vast range of groundwaters available in the region.

Rodent Pest Management

Well proven technologies have been standardised for rodent control with a view to minimise losses to grasses and standing crops.

Insect Pest Control

White grubs: Of the ecological and biological studies made on 25 species of white grubs, *holotrichia consanguinea* is the most important and predominant and serious pest. Chemical control studies against *holotrichia* spp. were taken up in some cultivators' fields in endemic areas. BHC 10 per cent dust at the rate of 100 kg/ha and sevidol at the rate of 25 kg/ha gave the best results by reducing the grub population 70 to 80 per cent and this approach for control of grub has received national recognition.

Termites: Fourteen species of termites have been observed as serious pests of crops, trees and grasses in desert and some of the salient control measures are given below:

(i) Control of forestry termites: 0.03 per cent of Aldrin EC when applied in a shallow pit around the base of trunk at the rate of one litre/plant gave full protection to the trees for at least one year.

(ii) Control of wheat termites: The wheat crop which is highly susceptible to termite attack can be safely protected by applying 5 per cent Aldrin or 10 per cent BHC dust at the rate of 10 kg a. i. per ha against termites.

Grasshoppers: Ecological and biological studies were conducted on grasshoppers affecting rangelands and grasslands in western Rajasthan. Grasshopper population was most abundant during monsoon season. The density of insects was significantly more in the overgrazed than ungrazed areas.

Dusting of BHC 10 per cent or Aldrin 5 per cent at the rate of 5 kg a. i. per ha was found very effective against yound hoppers. Adults on crop can be checked by spraying 0.8 per cent E.C. of Malathion or Carbaryl.

Katra (*Amsacta sp.*): Dusting carbaryl 5 per cent at the rate of 5 kg per ha methyl parathien (folidoe) 2 per cent at the rate of 1 kg per ha preferably in early instar stage reduces the damage effectively.

Integrated Natural Resources Survey

Since its inception, the CAZRI has surveyed over 9,000 km^2 area using a standardised base on aerial-photo interpretation aided by ground survey. Lately, the data is integrated in the from of Major Land Resources Unit (MLRU) concept based on homogeneity of biotic and abiotic land attributes. The information is assessed for resource potential and specific treatments are suggested for upgrading production on sustained basis. Various agencies of the state governments concerned have found the information very useful and special request surveys were carried out in Karnataka, Gujarat and Haryana besides Rajasthan. This appreciation is also reflected in the recent request from the Rajasthan government for an integrated basic resource survey of the Upper Luni Basin, for which the state concerned has also provided a generous grant of nearly Rs. 20 lakhs.

Groundwater survey for locating drinking water for Nagaur, Sojat, Bali and Falna towns at the request of Rajasthan State Public Health Engineering Department was done. Such survey was done also for Rampura-Mathania area to meet the drinking water requirement of Jodhpur town for the Rajasthan Ground Water Department. Investigations to locate groundwater were made for some of ICAR institutes, University of Jodhpur and defence organizations.

Geomorphic surveys have shown the existence of a prior drainage system in the Rajasthan desert. Courses of these channels have been reconstructed for surveyed area and it has been seen that these still regulate sub-terranean flow. Because of this, it makes potential source for water exploitation to meet the domestic and irrigation requirements. This has helped also in establishing the origin of salt basins. These salt basins lie at the dead confluence of an erstwhile drainage system. This fundamental study has also helped in enunciating some principles of land management.

An agricultural atlas of Rajasthan consisting of 34 maps depicting physical and socio-economic aspects has been published. The atlas is helpful in delineating regional pattern of resources in context of development planning. A groundwater atlas consisting of 30 plates has been compiled. It shows that in Rajasthan only 58 per cent of the total groundwater potential has been tapped. Of the total water exploited, 87 per cent is used for irrigation, which accounts for 50 per cent of the total irrigated area in the state. Atlases of agro-demography and sheep ecology have also been completed.

Solar Energy Utilisation

Research work on some of the important applications of solar energy for domestic, agricultural and industrial purposes such as heating water, drying and dehydration of fruits and vegetables, cooking and boiling of food and distillation of saline water has been carried out and some useful solar energy appliances have been developed as described below:

Solar Water Heater: A simple, low cost built-in storage type solar water heater has been designed and fabricated for the rural as well as urban areas. Five years of continuous and extensive testing of this solar water heater has clearly revealed that it can supply 90 liters of water of 50° to 60°C in winter months and warm water of 30° to 40°C in next day morning, by using a reflector-cum-insulation cover. The cost of water heater with insulation cover is Rs. 450 only and the cost per unit of thermal energy obtained with this water heater is nine paise per KWH. This water heater is being licenced for commercial exploitation through NRDC, New Delhi.

Solar Dryer: A solar cabinet dryer has been developed for dehydrating fruits and vegetables. With this dryer 15 kg of dates were dehydrated and 15 kg of grapes were converted into raisins within two to four days at Jodhpur even in monsoon season. Practically all types of fruits and vegetables can be dehydrated in this dryer. Chillies can also be dried in this dryer.

A forced convection type solar agricultural dryer for dehydrating 100 kgs of chillies in one day has also been designed and developed at this Institute.

Solar Cooker: Five different types of solar cookers have been fabricated and their performance compared in summer as well as in winter seasons at Jodhpur. Out of all these five cookers, the solar oven is found best. The total cost of this oven is about Rs. 350. This solar oven is suitable for cooking, baking and roasting the food material within 30 to 75 minutes under clear sky conditions. The main advantage of the solar oven is that its performance is not affected by the wind and there are no chances of dust fall in the cooking pot which are the problems in the arid zone of Rajasthan. Moreover, the food remains warm if kept inside the oven for hours together even after sunset.

Socio-economic Factors

Man has been one of the most important and active agent of botic interference in the fragile ecosystem of arid zone of Rajasthan. Human resources assessment, therefore, formed an important component of integrated natural resource surveys for the resource use planning. An important social problem of the arid region is of the nomads, historical, political and cultural factors combined life. The nomads in the present day prove a menace for the whole society and their sedentarisation is inescapable. The rehabilitation measures formulated on the basis of their kinship structure and cultural values are likely to yield fruitful results. The socio-economic surveys undertaken among the settled population provide details about the population characteristics, social and economic correlation of various caste groups, households structure, rural working force, class of farmers and holding size, animal-human-vegetation-relationship, form of settlement, historical perspectives of land use, cropping pattern, rural crafts, indebtedness, etc. Intensive studies on socio-economic aspects related to desert spread and desertification in the arid zone of Rajasthan included study of biotic factors and the macro parameters. The opinion polls on desertification were employed to delineate the factors effecting desertification.

2

Social Aspects of Desert Development

S.P. MALHOTRA

Human population has been living in balance with the natural environment in the arid zone of India since centuries, the balance being maintained sometimes by natural and at others by the adjustment mechanisms adopted by the desert inhabitants. History has it on record that man not only lived in this desert since ancient times but great civilisations flourished in the region. Studies conducted on the Indian desert have revealed that ecological deterioration and accentuation of desertification have been occurring in the region in recent times due to the intense human livestock pressure. The portents rather than the promises currently weigh heavily in any programme of the development for the desert areas. Economic, social and political considerations condition the multifarious problems involved in developing the arid areas. The problems of economic development are conditioned by the physical characteristics, the hot Indian arid zone being characterised by low and erratic precipitation, frequent occurrence of drought, high temperature with high wind velocity resulting in severe kind of wind erosion hazards. The land forms comprise barren rocky stretches, vast sand dune areas and undulating topography. Drinking water is scarce. The sub-soil water level is often too low and has high salt content. The economic problems have also a human dimension and most of them are embedded in the social structure of the population inhabiting the desert areas. By arid zone standards, the Indian zone is most thickly populated. In the process of eking out existence more and more marginal lands have been exploited to effect accentuation in desertification in the region. There also has been a breakdown in the traditional social fabric so well knitted for the balanced use of the ecosystem. Any programme of arid zone development must, therefore, take into account the human factor for diagnosing the processes by which resources are being depleted, for assessing the nature and extent of present ills and for indicating the remedies.

Traditional System of Resource Use

Resource use being the decisive factor of the pattern of their living, the population of Rajasthan's arid zone divided itself into two major groups, viz., the nomadic and sedentary, to create one integrated unit and provided for the judicious use of the ecosystem. Each type of the two populations was further broadly grouped on the basis of their traditional occupations. There was a definite division of labour, a pattern of exchange relationships and distributive use of the ecosystem within nomadic groups, between the nomadic and sedentary populations and within the sedentary populations. The functional specialisation of different castes and communities not only knitted them together through relationship of exchange of produce and services but also resulted in the distributed use of the ecosystem. Within this broader framework, both the nomadic and the sedentary population had further evolved mechanisms for the maintenance of resource use equilibrium.

Land and Resource Utilisation by Sedentary Population

The sources of livelihood and the way of life of the sedentary population had been conditioned by the disturbances in the ecological balance owing to the vagaries of the climate. Agriculture in the arid areas has always been a big gamble and the droughts and famines have been of common occurrence in the region. The desert dwellers had, therefore, evolved tenacious means and adjustment mechanisms to cope up with such situations. Being fully aware of the dangers of getting wiped out due to high mortality specifically during famine periods, the arid zone society had built-in high fertility norms in their culture. Social values were predisposed for more children and positive sanctions for fertility far outweighed the negative ones. Early marriage and begetting of children were integral parts of the social ethos of these people. Any deviation from the established norms was looked down upon as an aberration and as wholly incompatible with the social fabric of the society. Divorce was a rarity, if not altogether unknown, and widowhood soon culminated in remarriage.

Since the rainfall would not only vary five to ten folds from one year to another but also during the same year at places just a few kilometers from each other, the farmers were never sure of harvesting grains. Most of them, therefore, practiced a subsidiary occupation of livestock raising which ensured survival of the society. The size of the land held per household was quite large and was scattered far and wide within several fragments in the boundary of the village. The farmers were generally of the opinion that greater number of fragments of land provided surety for getting some harvesting from some plots as they may get rains while others may remain absolutely dry due to the erratic nature of the rainfall.

The above system thus had an in-built provision of leaving the lands fallow during the periods when these did not get rain. Even otherwise also, the farmers had developed the practice of leaving the land fallow after four to five years of its cultivation. This resulted in the land regaining fertility. The fallow lands also provided grazing resources for the livestock which, in turn, provided manure for these fields.

The system of cultivation by rotating and following a proper mix of different crops further provided security in this risk prone agriculture. After fallowing the land, usually guar (*Cyamopsis tetragonoloba*) was grown in the first year and the land was ploughed thrice. This was followed by a mixed crop of bajra (*Ponnisetum amerecanum*), moth (*Vigna aconitifolia*), moong (*Vigna radiate*) and till (*Sesamum indicum*) in the ratio of 20:1:1:1. In the subsequent years, the proportion of moth crop was increased as this required lesser fertility.

The desert dwellers had evolved tenacious means to use the conserved moisture by way of cultivation of 'khadins', which essentially are natural depressions in the midst of rocky and stony terrains—characteristics of arid environments. From times immemorial, erosion in the catchments has been filling up these depressions with fine sediments resulting in the development of localised silty clay-loam and clay-loam spoil pockets, called 'khadins', in the otherwise sandy and rocky territory.

The 'khadins' are usually flooded during monsoon and if necessary, a bund is laid down with a view to collecting the runoff from the catchments. These low lying areas are cultivated as and when the water recedes and they become ploughable in the *rabi* (winter) season. In a few instances of lighter and better drained soil pockets, and also in cases where smaller catchments with lesser runoff are involved, 'khadins' are cultivated in the *kharif* (monsoon) season also.

Similarly, in view of the high incidence of mortality among the livestock, the arid zone society had developed a tradition of considering it prestigious to possess a large number of livestock. Quantity over quality was preferred as the periodic high mortality used to bring forth a balance between the livestock population and the grazing resources. To have effective and balanced use of the ecosystem, and for utilising different types of grasses, shrubs, tree, leaves, etc., the households used to herd a mixture of animal types including cattle, camel, sheep and goats etc. Only in very rare cases, the households resorted to raising only one species.

Since the majority of people resided in scattered home-steads and the man-land ratio was quite large, there was no difficulty in procuring fuel wood for its procurement did not amount to any degeneration of vegetation. There were, however, in-built customs for preservation of vegetation on certain types of lands and also for raising certain trees for religious reasons, e.g., it was religiously prohibited to cut any vegetation from the lands in the immediate vicinity of temples and religious places, known as oran lands. Similarly, women daily watered the peepal (*Ficus religosa*) tree regarding this as a pious act. This indirectly encouraged conservation of vegetation and raising of some trees providing good shade and protecting the soil.

The introduction of new social order and practices has led to intense human and livestock pressure in the arid zone. The high fertility norms continue while the incidence of mortality, specifically the infant mortality, has substantially declined. As a result, the Indian desert is one of the most thickly populated deserts of the world. In the arid zone of Rajasthan, starting with a base of roughly 3.567 millions in 1901, the population of the region registered a linear escalation and increased over three times (10.236

millions) than its size in 1901. As against 186 per cent growth rate of population (1901-71) in the arid zone of Rajasthan, the Rajasthan state, as a whole, registered an increase 150 per cent. The increase in the country during the period has been of the order of 132 per cent. This evidently reveals that the growth rate of population in the arid zone of Rajasthan has been relatively higher. The enormous acceleration of growth in recent times is exhibited by the fact that the number of new additions in only ten years (1961-71) worked out to 62.64 per cent of the number present at the beginning of the century.

Most of the labour force (over 80% of the working population) is tied up to a premodern agriculture which suffers from an additional handicap of increasing scarcity of crop land. This major form of human occupation is marked by instability that carry the hazards of sudden failure leading to famine conditions. The situation is more revealing when we take man-land ratio and the high proportion of labour force dependent on land. It is observed that while the total land available per household was 17.77 ha 14.69 ha and 12.40 ha during 1951, 1961 and 1971 respectively and only 7.52 ha is likely to be available by the end of the century. Similarly, total cultivable land available per household reduced from 13.72 ha to 9.95 ha in 1971 and only 6.33 ha is likely to be available in 2001. The present broad based age pyramid and civil conditions of the population exhibit an expansive future growth potential.

Landuse statistics revealed that as a consequence of high growth rate of population, marginal and sub-marginal lands have been brought under agriculture. The percentage of net sown area in arid zone of Rajasthan had been 26.1 in 1951, 39.0 in 1961 and 42.2 in 1971 while the percentage of double cropped area increased by 50.3 per cent during 1951-61 and by an additional 8.34 per cent during 1961-71 and the area under less intensive uses of land (barren, cultivable and uncultivable waste lands, permanent pasture and fallow lands) declined by 18.9 per cent during 1951-61 and further by 6.2 per cent during 1961-71. Thus, during 1951-71, there had been a substantial shrinkage of grazing lands due to extension of cultivation on marginal lands. One factor responsible for the cropping of more and more marginal land is the adoption of short term welfare measures for the growing population rather than a long-term measure to ensure the capability of land. Land distribution policies in the region have aggravated rather than ameliorated the maladjustments in the landuse pattern. Most land allotted to the landless marginal farmers and others were marginal and sub-marginal lands which were put under cultivation. The yield per unit of cultivated area is declining. The annual growth rate, from 1954 to 1970, of principal crops shows a negative trend: millets (–) 0.66, sorghum (–) 4.88, pulses (–) 2.17, sesamum (–) 6.08, gram (–) 1.20 and barley (–) 0.18.

Paradoxical, as it may appear, on the one hand, there had been a shrinkage in area under grazing lands and livestock population increased from 9.4 millions in 1951 to 14.4 millions in 1961(53.2% increase) and 18.0 millions in 1971 (25% increase) in the arid zone of Rajasthan. The density of livestock per hundred hectare of grazing lands rose from 72 in 1951 and to 175 to 1971. The continuous increase in the livestock population is a serious threat to the forage production. Considering the present carrying capacity of the grazing lands, there is acute overcrowding of livestock on grazing lands which evidently presents a sad picture and points to the possibilities of further deterioration of grazing lands. Overuse and over-exploitation of the grazing land is further evidenced by taking into account the total production potential of forage and its requirement during different periods. The forage production was adequate in 1951 and it exhibited deficit during all the quinquennial period which rose to the extent of 31.8 per cent of its requirement in 1971 from a slight surplus in 1951.

The animals in arid Rajasthan thus partially starve during to year and to eke out maximum to satisfy hunger to overuse of grazing land goes on to add to the impoverishment of land vegetations.

The over-exploitation of over-dwindling vegetational resource that started ever since the historical period, was accentuated during the times of princely states and is on the verge of doom at present in the arid land of Rajasthan. Due to overcultivation, the feedback regeneration process of trees and shrubs is being hampered by the cultivator, for he maintains a bare minimum tree-shrubs population in the crop field, and rest is uprooted, resulting in lowering the woody biomass production.

The tree and shrubs are indiscriminately cut not only from their own fields but from government and other unoccupied lands for: (i) meeting the fuel requirement; (ii) the top-feed of livestock; (iii) thorn fencing; and (iv) construction of thatched hutments.

The fuel energy requirements per person per annum was calculated by National Council of Applied Economic Research to be 0.254 tonne of coal equivalent.

			(million tonnes)
Exploitation of Woody Biomass	**1951**	**1961**	**1971**
Fuelwood requirements	1.64	2.39	3.02
Wood for other requirements	0.21	0.28	0.31
Total wood produce standing on ground	26.04	23.48	20.76

These estimates clearly indicate that the utilisation trend of wood is showing a gradually increasing trend whereas the total wood produce is showing a declining trend. The requirement and overincreasing demand is huge and overexploitation of the vegetal resources is inescapable. The older people in the villages admit that during the course of years the number of trees and shrubs has gone down steadily and not even 35-40 per cent of what was available thirty years ago has been left over. Additionally, a large proportion of population burns dung which would have otherwise been used for manure purpose. Thus, the rapidly growing population has inflated the demand for different purposes while it has reduced the area under fallows, waste and forests.

Other Sources of Livelihood

The main sources of livelihood, other than agriculture and animal husbandry, are agricultural and casual labour and the traditional caste occupations like carpentry, blacksmithy, oil pressing, drumming, dyeing, money lending, shopkeeping, pottery making, basket making, leather work etc. These are generally followed as subsidiary occupations and generally cater to the needs of the village. Since there is at present overcrowding in agriculture, it would be necessary to develop cottage industries to provide the much needed supplementary income.

Indebtedness

In the desert regions, more often than not, the crops fail and the savings accumulated over a number of years are not enough for meeting the future capital needs of investment in different occupations and for meeting the heavy expenses to be incurred in socio-religious ceremonies. The survey conducted in the central and lower Luni Basin revealed that six-tenths of the rural settled households were indebted. The average indebtedness per household comes to Rs. 533. Most of the loans have been procured for socio-religious ceremonies where obligations to community over-ride prudence and force the cultivators into the clutches of the *Mahajan,* the traditional moneylender, who still remains as the chief source of credit and quite a substantial part of the debt is the high rate of compound interest on the loans charged by him.

Nomads and Nomadism

Famines of different kinds, namely, *akal* (great famine), *jalkal* (scarcity of water), *tinkal* (scarcity of fodder) and *trikal* (scarcity of fodder, water and grain) have been of common occurrence in the region. To cope up with the such unfriendly climatic and environmental conditions, the inhabitants have been resorting to nomadic and semi-nomadic life. They are not a separate ethnic group and do not have a separate territory with exclusive right and economic framework. The resource use being the decisive factor of the pattern of their living, the nomadic groups of the arid zone may be broadly grouped into four categories, viz., (a) the pastoral nomads (raikas, Sindhis, barihars, billochs, etc.); (b) the trading nomad (banjaras, ghattiwala jogis and gawarieas); (c) artisan nomad (the gadoliya lohars, sansis and sattias); and (d) miscellaneous types of nomads (nats, kalbeliya jogis).

Relationship with Settled Population

Each type of nomad is associated with some kind of livestock which makes indiscriminate use of the meagre available water and grazing resources and destroy the local soil-conservation measures. The nomad in the present times, thus, are a menace to the whole society and their sedentarisation is imperative. The opening up of the means of communication has reduced the importance of distribution activities by the trading nomads. The shrinkage of grazing lands in some areas due to the extension of cultivation has created difficulties for the cattle-breeding nomads. Villagers are no longer dependent upon those nomads who once rendered specialised services. It is evident that the sedentary population, in general, does not welcome the nomads to visit their areas.

It was thought that the settled population did not get any benefit from such nomads as *sansis, sattias* and *nats*. About the visits of *banjaras*, two-third of the respondents felt that there was no benefit, whereas one-third of them felt that the settled population could purchase salt and other commodities from them. About the other trading nomads, a significantly greater percentage felt that the articles could be purchased, sold, exchanged or repaired.

Pastoral Nomads

The pastoral people inhabiting the Rajasthan desert have been carrying on livestock breeding for generations and have contributed to the economy of the region by way of providing milch cattle, *ghee*, wool, mutton, etc. A detailed survey conducted among the livestock breeders of the Anupgarh-Pugal region revealed that these nomads had been facing difficulties in grazing their livestock as there had been a shrinkage in the extent of grazing lands owing to the extension of cultivation and also with the coming of the Rajasthan Canal to this area, the economy of these livestock breeders is being much disturbed. The inhabitants of the region are not completely nomadic, but have permanent homes and generally have parcels of land. Their nomadism is directly dependent on the necessities of their herds. The economy of these breeders is largely dependent on the harvesting of the available water and the efficient use of extensive grazing lands available in the area, as the cultivated area is only 4 per cent of the total land surface.

Among the breeders, water harvesting for livestock have been in vogue for generations. Having evolved a regional ecological association, the traditional livestock breeders of the region are maintaining it more profitably and effectively for raising livestock of the region. This old method, presently known as *toba*, constructed amidst the sand dunes which provide water and pasturage for the livestock is a very fascinating place to live in. A hard plot of land with very low porosity is searched for constructing the *toba*, and again this place should form a good catchment area, so that water from all the four outsides runs and collects at this place. Each village of the cattle breeders has 5 to 6 *tobas* dug, depending upon the caste and community composition.

With the onset of the monsoon, there is a dispersal of herds far and wide within the boundaries of the village, wherever surface water is available in the dug-out *tobas* which are constructed and kept in repair for the efficient use of the meagre available water and grasses. The use of *tobas* is regulated by conventions and it is customary to make use of other *tobas* when the water in one's *toba* is exhausted. Nobody was thus prepared to deepen his *toba* lest the grazing resources are utilised by others. Thus the people stressed that the *tobas* had to be dug deep on a large scale rather than making any piecemeal efforts. In case the rains fail, the breeders migrate with their livestock to the irrigated areas of the Ganganagar and Ferozepur districts. Before the formation of Pakistan, they used to go upto the Bhawalpur region which is now a closed boundary.

It has been recommended that each household in the canal commanded area may be allotted 6.3 ha of land and a partial mechanisation of cultivation may be resorted to. Dairying may be introduced as an industry on scientific lines. Sheep-raising is the chief subsidiary occupation in the region. At present, the sheep-owners have to travel long distances and their nomadism can be arrested by upgrading the short grass rangelands by reseeding these pastures with grasses, etg. *anjan* (Cenchrus ciliaris) and *dhaman* (Cenchrus setigerus). The introduction of the mutton industry and the setting up of wool-grading and shearing centres shall further save a lot of their energy and fetch the breeders much higher income than they earn at present.

Artisan Nomads

The second important category of nomads is constituted by those who specialise in certain trades, like the much-publicised *gadoliya lohars*, sansis and sattias. The sattias and the sansis carry on a limited trade in cattle and render veterinary services to the livestock of the villages on their routes. It is the nomads like these who are known for their immoral activities, illegal transaction and skill in thieving and who have created a resentment in the minds of the settled population against the hapless nomads. They are placed even by the nomads themselves in the lowest category of their social hierarchy.

The nomadic blacksmiths are locally called *gadoliya lohars*. They are one of the artisan nomads in the states of Rajasthan, Punjab and Madhya Pradesh. Their territorial extension is even in the semi-arid and humid environment. The traditional occupation of the *gadoliya lohars* is blacksmithy and trade in cattle. Their movements to different places are dependent upon the availability of blacksmithy work and prospects of trade in cattle. They have developed a symbiotic relationship with the sedentary farmers for supplying them with agricultural tools and implements. They also sell utensils for the farmers' households. These materials are supplied to the rural population on cash or exchange basis. The technology of manufacturing the tools, implements and utensils of iron is indigenous and it involves hard work and a large amount of labour. Almost all the members of a family remain engaged in this occupation. On an average, a household earns about one thousand rupees annually from this occupation. Uncertainty of the availability of work and risk in earnings has created feelings of fraternity and mutual give-and-take in society. Unlike other nomadic groups, generally they do not take loans from the moneylenders. They are economically and culturally very backward. Their children are normally not sent to school, and as such, illiteracy is common.

The nomads move about in small kinship bands in their bullock-carts with their families. The nature of movements of different bands of these artisan nomads follow a regular cycle on regular monthly intervals. They owe their allegiance to a definite demarcated area. In one band, the number varies from one to twelve families. During movements, they split themselves into smaller constituents which group and regroup at different places, depending upon their social and economic needs. Their movements are limited only to those places which are connected by roads or tracks on which their bullock-carts can go. Owing to the recant changes in economic and political set-ups and the increasing means of communication, the *gadoliya lohars* are not as welcome in the villages now as about 15 years ago. About one-fourth of the *gadoliya lohars* stated that there has been a definite decrease in their getting work of blacksmithy and in cattle trade in recent years. Ninety-five per cent of the household surveyed, therefore, desired to abandon nomadic life and lead a sedentary or semi-sedentary life, leading finally to sedentarisation.

Efforts have been made by the Government of Rajasthan to sedentarise *gadoliya lohars in* colonies consisting of many households at one place, but such sedentarisation has achieved only a partial success. Most of them desired to sedentarise in small kinship groups consisting mostly of agnatic (related on the father's side) and matrimonial relation. They desired sedentarisation in a scattered fashion at a central place within the present area of their movements, locally called by them *chokla*.

Such a settlement shall ensure the continuation of their symbiotic relationship with the sedentary population, the provision of marketing facilities for the articles fabricated by them and shall also provide them with better prospects for being allotted suitable agricultural lands, which they are most desirous to get. Over and above, this social change shall also rule out the occurrence of any feuds which are likely when too many households of different clans are settled at one place.

Trading Nomads

The third important category of nomads—designated as the trading nomads, consist of the *banjaras, ghattiwala jogis* and *gawariyas*. The *gawariyas* mainly cater for the village women, selling beads, bangles and trinkets, and use the donkey as the transporting animal for a fixed area of movements. Like that of the *gawariyas*, the area of the *ghattiwala jogis*, who manufacture and trade in grinding-wheels and earthen smoking-pipe is also fixed. While some of these trading nomads are on the move perpetually, others lead a restricted settled life, taking to cultivation or working as daily wage agricultural labourers during the cultivation season. After the season, they start peddling their age-old wares again.

The *banjaras* are one of the most publicised trading groups of nomads. They are organised into a distinct class in different territories. Each clan is divided into smaller groups called *tandas*. In each *tanda* there are six to twenty families. The doctrine of collective responsibility operates among the members of a *tanda*. The families within the *tanda* are bound by the ties of kinship which give security to the members. The rights, obligations and expectancies towards one another are well established. The structure of the family is patriarchal. Early marriage is very common among them. Every *tanda* has a headman who commands respect from other members.

The main occupations of the *banjaras* is trading in salt. They roam from village to village with bullocks which carry goods and other material possession that are only a few and limited to the bare necessities of life. Even though nowadays the salt trade is carried on by big wholesale traders with modern transport facilities, it remains the main source of sustenance of most of the families of the *banjaras*. The trade is carried on with the family labour force. The women render only minor assistance in economic pursuits.

With the onset of rains, the *banjaras* move towards saline basins where salt is manufactured. After rains, grass and water are available on route for their animals. In addition to the salt trade, they sell Fuller's earth and onions in the remote village of the desert tracts. These *banjaras* move towards the Gujarat state during the post-monsoon period. There they get ready employment in the form of hauling of commodities, e.g., building materials and grains by using their bullocks.

The nomadic *banjaras* do not avail themselves of any community facility and have no social and cultural relationships with the settled population whom they meet only for trade purpose. There is some communication between the money-lenders and the *banjaras*, as the latter generally take loans from them for their trade.

The *banjaras* are economically backward and their standard of living is very poor. As a result of improved means of communication and the opening up of hitherto inaccessible areas to road transport, their traditional relationship of mutual dependence on the sedentary population has broken down. The *banjaras* find it difficult to sell their commodities with a good margin of profit as in the past. The shortage of grazing land is another problem for them. Their earnings have declined and several of them are in perpetual debt. For sedentarising the trading nomads (*banjaras*), the *tanda* should be taken as one unit for settlement purposes in different villages. It shall be highly desirable to make use of the existing administrative machinery among the nomads rather than supersede it.

The fourth class consists of miscellaneous nomads, e.g., the *nats* and *kalbeliya jogis*, who earn their livelihood by showing acrobatics, jugglery and snake-charming at village fairs and at roadside gatherings.

The technological, economic and political changes of the recent times are influencing the nomadic population in this region with such rapidity that they are facing a social decay. This influence is chaotic insofar as it is not based on balanced and well-conceived plans, and is not adaptable to the traditional nomadic way of life, and their aptitudes. The sedentarisation of the nomadic way of life is gradual and a slow process. Though all-round development of the region may result in a spontaneous sedentarisation of the nomads, there is a possibility of social as well as cultural inhabitation of the nomads. The question that haunts the ecologists and sociologists today is how to absorb these people within the framework of modern living and make them useful members of society by exploiting their natural aptitudes.

Conclusion

To combat the process of desertification the foremost necessity is to impose checks on the relatively high increasing population in the arid zone of Rajasthan. High fertility norms and social values predisposed for having more children need to be changed to reduce fertility. It may be essential to have an intensive population education programme both in and out of the formal school system because of the high rate of school dropouts. The norms of raising the marriage age both for boys and girls and of 2 to 3 children per family should form a part of the value system.

The traditional system of cultural integrity and resource use has largely broken down. There is increasing overdependence of the population on agriculture and the man-land ratio is fast declining. It is therefore necessary to develop avenues of employment other than agriculture and also to encourage and make use of the traditional institutions rather than to truncate the same.

The practicability of extending improved practices needs to be thoroughly scrutinised. It will be useful to encourage as much participation and involvement of farmers in various activities as this predisposes the farmers towards adoption of innovation. Providing full technical knowhow of the innovation is another important factor.

Animal husbandry practices are still largely traditional. The produce is meagre and the sale of livestock and livestock produce is extremely limited. The decidedly greater advantage of keeping a smaller number of animals of superior breed are understood by the farmer. It is, therefore, suggested that facilities for exchange of poor quality livestock with better ones be provided. This should include giving of bullocks in exchange of bulls as due to prejudice against castration, the exchange of bulls for bullocks done by the villagers at present are to his disadvantage. The Panchayat Samitis should take upon themselves the responsibility of maintaining bulls of better breeds. The useless livestock should be sent to *Goshalas*.

Vigorous efforts need to be made to control diseases dissemination, to demonstrate improved techniques of animal husbandry and to extent supply, marketing and other facilities.

Development of farm forestry will take place if cultivator, in the arid areas are encouraged to plant a few recommended tree species around their habitations, sand dunes, pasture lands, road sides and near the water sources. Religious feelings of women folk should be exploited by having a small grove of peepal (*Ficus religious*). Again in the *orans* (lands attached to the religious places) cutting of vegetation is prohibited and important trees, bush and grass materials should be planted and propagated at such sites.

The present cottage industries of shoe making, basket making and weaving need to be reorganised. Credit facilities should be set up which will buy all the products and acquaint the industry with the types of product required. Quality needs to be enforced and training facilities provided. It will provide the additional opportunities of employment to the villagers by setting up of rural estates of small industries.

Cooperative societies have been functioning in some villages but these have been unable to replace the moneylender. One of the chief reason is that these societies give loans only for productive purposes and not for other purposes including the socio-religious needs of the community for which people more often than not run to the counting house of the Mahajans. Also the foresight of the people for advantage of controlled credit and its punctual repayment is lacking. The propaganda and persuasion to induce such a foresight among the inhabitants of the region should be take up by the extension workers without delays.

Some of the recommendations based on our studies on the nomads are provided below:

1. For the area coming under canal irrigation it is recommended that each household of livestock breeders may be allotted about one square (6.3 ha) of land and partial mechanisation of cultivation may be resorted to. However, about 800 households in this region who have large number of cows of *Rathi* breed should preferably be sedentarised in colonies and dairy industry be introduced among them on scientific lines.

2. The system of rotational grazing and management of pasture lands and reseeding of vast fallow land with good strains of grasses should be introduced.
3. Cooperative forage collection centres and fodder banks should be opened.
4. Improved animal husbandry practices should be introduced on a large scale in the region.
5. Introduction and development of cottage industries of spinning and weaving and mutton production should be taken up.
6. It may be essential to sedentarise these nomads in bands of small kinship groups within their own *chokla* or present area of movement as is desired by them. Such a settlement shall ensure the continuation of their symbiotic relationship with the sedentary population, provision of marketing facilities for the articles fabricated by them and shall also provide better prospects for allotting them suitable agricultural lands which the *gadoliya lohars* are most desirous to get. Over and above this shall also rule out occurrence of any serious feuds which are likely to occur when too many households of different clans are settled at one place.
7. For sedentarising the trading nomads (*banjaras*) the *tanda* should be taken as one unit for settlement purposes in different villages. It shall be highly desirable to make use of the existing administrative machinery among the nomads rather than supersede them.
8. Suitable arrangements for storage of drinking water, school-cum-community centres and marketing of surplus produce should be made for all the nomads.
9. The cooperative societies should be formed for giving loan on nominal interest for their efficient running of their traditional occupation of blacksmithy and cattle trade as well as for the proper initiation and accelerated adoption of cultivation.

3

Monitoring of Desertification

K.A. SHANKARNARAYAN

Desertification means a sustained decline in the yield of useful crops from a dry area accompanying certain kinds of environmental change both natural and induced. Earlier man had played a relatively passive role of mere existence as part of nature. But now man has completely transformed the very earth into a source of sustenance and exploitation to meet his own ends. This has led to an unprecedented proliferation of his species and exponentially increasing problems arising out of excessive resource utilisation and interference with the ecosystem. As a result, at the turning point of the twentieth century, there is serious thinking at various levels on the utter necessity of calling a halt to indiscriminate exploitation of nature and making a start towards developing a symbiotic relationship with nature and environment. It is in this context of crisis and the possibility of encountering a "limit to human growth" that the United Nations General Assembly, at its twenty-ninth session in 1974, decided to initiate concerted international action to combat desertification and convened a United Nations Conference on Desertification during August-September 1977 at Nairobi to give impetus to international action to combat desertification.

Impact of Exploitation

Most ecosystems can be cropped in some way without endangering their powers of recovery. Because a surplus is provided by some species as a means of ensuring their resilience after severe stress, the crop is often large in good times but is very small in bad. In a more modified ecosystem, as in agriculture, much greater productivity of useful crops is bought at the price of labour, technology and imported materials.

At some level, exploitation ceases to benefit and begins to degrade productive potential. Degradation can be either chronic or acute. In pastoral production desirable species may be slowly eliminated as their competitive ability is sapped by grazing and less valuable plant species may thrive. In agriculture, there may be a gradual loss of nutrients as they are lost from the farm in crops sold elsewhere. Pastoralism, rainfed agriculture and land cleared for industry or housing can all suffer the same kind of acute loss when rare heavy showers or strong dry winds erode in hours soil that has taken centuries to accumulate. Irrigated land suffers its own type of desertification. Each of these processes lends its peculiar imprint to the pattern of desertification.

Evidences of Desertification

In the context of global concern on desertification expressed by UN General Assembly and the decision of UNEP to provide impetus for a plan of international action to combat desertification, a case study on desertification in Luni block of western Rajasthan covering an area of 1900 sq. km. was undertaken in 1976. Some of the salient findings on the process of desertification in time scale in terms of geomorphological, geohydrological, soils and soil fertility and vegetation are presented below.

Geomorphological Evidence of Desertification

The landforms of the Luni development block were formed by both the fluvial and aeolian geomorphic processes operating on various lithological formation, and at least under two major humid and two major arid phases during the late quaternary period. As a result of those climatic changes and subsequent human activities, twelve major landform units were formed in the block, which have also been termed as the ecosystems. A comparative study of the changes in these ecosystems reveals that prior to 1958, the age-old cultivation practices along the slopes of the stabilised dunes and in the light to medium textured flat aggraded older alluvial plains around the dunes, and drought condition prevailing from time to time had favoured deflation and deposition of sediments to such an extent that a new ecosystem, viz, sandy undulating aggraded older alluvial plain comprising of an area of 52,430 hectares with longitudinal and transverse and ridges of 3 to 6 m height, sandy hummocks of 90 cm to 1 m height and sand sheets of 20 to 200 cm thickness, was formed. Some parts of this ecosystem, particularly around Gangana, Bhakrasni, Doli, Jhanwar, Hingola, Rabaria, Uter, Kagnada, Bhetnada and Phinch villages were affected by sand deposition hazards, while some other parts, mostly in the west, were subjected to sand deflation hazards. In the medium to heavy textured flat aggraded older alluvial plains construction of tanks and the Summair Samand Canal across the prior channels of the Agaria Nala created ponded situation, resulting into the rise of the sub-surface brine level, salinisation of the tanks and canal beds spread of salinity in the nearby agricultural fields, especially around Lolawas, Bhetnada, Kharabera Patarobas and Mori Sutran. A total of 830 hectares land of the block were affected prior to 1958 by such man-induced salinity. Accelerated fluvial erosion in the forms of sheet-wash, rills and gullies had affected the piedmont plains and gravelly pediments.

Since 1958, there have been no significant increase in the zone of sand deflation and deposition in the sandy undulating aggraded older alluvial plains, and deposition. But exploitation of groundwater and irrigation in Jhanwar-Doli-Boranara sector of it have stabilised the sands and increased the productivity, which indicate a sign of de-desertification. Exploitation of groundwater along the younger alluvial plains of the Mitri has also improved the condition there, but salinity has spread considerably in the medium to heavy textured older alluvial plain due to the construction of more tanks along the prior channels of the Agaria Mala. A total of 1560 ha of land of this plain from Nalkosni to Nimbla has become saline since 1958. Salinity has also spread at Shikarpura and Kankani in the younger alluvial plain of the Luni due to constant irrigation with saline well-water. The wells have become saline after the construction of a dam across the Luni at Pichak near the eastern border of Jodhpur district. Construction of the dam had created favourable geomorphic condition for the development and spread of salinity hazard in the surrounding areas. Then in less than 100 years time, the salts began to be leached downstream along the Luni and gradually due to the rise in water level, turned the sweet water wells along the river saline. Salinity of wells have spread now upto Dunara, downstream of Shikarpura. At Shikarpura and Kankani the farmer used the saline well water for irrigating the salinised 830 hectares of younger alluvial plain.

Based on these findings, it has been worked out that in future about 16330 hectares of lands in the sandy undulating aggraded older alluvial plains and 1770 hectares of lands in the light to medium textured flat aggraded older alluvial plains may be affected by sand deflation and depositional hazards and 4030 hectares of land in medium to heavy textured flat aggraded older alluvial plain may be affected by salinity hazard in the near future, if some rational and scientific land utilisation methods are not adopted immediately. The rest of the area of the block does not seem to affected by such desertification process in the near future. It has also been concluded that at present there is no evidence of climatic desertification from the block. But intensive biotic activities have caused conspicuous changes in the geomorphological properties of the different ecosystems, leading to desertification.

Geohydrological Evidence of Desertification

Surface Water Resources (Tanks/Nadis)

The water in these *nadis* generally last from one to 10 months in a year. According to a survey the water in the *nadis* lasted for less than three months in 15 per cent cases, 3 to 6 months in 68.9 per cent cases, 6 to 9 months in 11.8 per cent cases and more than nine months in 4.3 per cent cases. During 1976, it was observed that in 35.2 per cent cases, the water lasted for longer duration than observed in 1966. There could be two reasons. Firstly, 1974-75 was heavy rainfall period and secondly, some dredging activity was undertaken for these *nadis*. In 29.4 per cent cases, the capacity of the tanks have been found to have reduced than previously observed. Such tanks are generally located in areas of higher biotic activities and in sandy environments. This indicates that due to erosion and movement of sand, these tanks are silting up. In a few cases like Jhanwar, after the discovery of groundwater potential area, people have paid less attention to the maintenance of *nadis*. As a result the capacity has reduced. In 35.4 per cent cases, which are generally located on piedmonts and older alluvial plains, capacities of *nadis* have not been much affected within the period under study.

Present Utilisation of Groundwater

Generally the groundwater in the study area is being exploited for human and cattle requirements and minor irrigation. The irrigation is practiced on very small areas which are generally confined to the river zones of Mitri and Luni river. In the recent past a groundwater potential basin has been discovered in Pal, Jhanwar and Doli areas, where irrigation of crops is developing at a fast speed.

Pal, Jhanwar and Doli Areas

This basin is composed of sand and gravels overlying piedmont surface of cemented Vindhyan sandstone. It extends in about 25 km area in the form of a curvilinear band 6.4 km wide and has saturated thickness between 24 to 30 m. which thins out at Doli. The static water level ranges at 35-40 m below ground level, and has a total storage of 77 mcm. This basin is recharged during monsoon season by infiltration and runoff originating in the hilly region just north of the area. The area has an average of 359.8 mm rainfall, about 90 per cent of which occurs during the months of monsoon. Statistical analysis indicates that this area receives 559 mm surplus rainfall in every five years and approximately 663 mm surplus rainfall in every ten years. A major portion of this surplus rainfall goes away as runoff through Luni river (870 mcm), leaving behind a very small proportion for recharging the aquifers. The safe yield of this basin has been estimated at 4.85 mcm per year of good quality i.e., with total soluble salts contain upto 500 ppm. But it could be upto 7.5 mcm per year of poor quality of water. The draft from this aquifer has increased of the order of 5.0 mcm per year within the past few years. Though the water table conditions (which has come up during 1974-75) does not apparently show overdraft, but the above stated facts clearly indicate that this area has meagre groundwater potentials and should be utilised with caution, specially for irrigation of crops.

Exploitation of Groundwater Resources

To study the areas, the following classification has been adopted:

Unexploited : These are the areas where no attempt has been made to use groundwater.

Under-exploited	:	These are the areas where wells can stand draft for over 12 hours per day with common type of water lifting devices, i.e. persian wheel, mote and bull (ad. discharge 2 to 3 lit/sec.)
Moderately exploited	:	In these areas the wells will stand draft for 4-12 hours in a day.
Over-exploited	:	In these areas the wells get exhausted within 4 hours of its operation and has to be left free for 20 hours for recharge.

The comparative results of 1966 and 1976 investigations are shown below:

	1966	*1976*
Unexploited	71.8	—
Under-exploited	4.2	11.6
Moderately exploited	17.8	11.6
Over-exploited	6.2	76.8

The above figures very clearly indicate that the groundwater resources are being depleted rapidly.

Theoretically working out, the area has a total recoverable storage of 219 mcm. The average rise in water table between the pre and post monsoon period during normal year of rainfall is 0.8 m.

According to the composition of aquifer safe yield can be taken 3 to 5 per cent. This gives an annual recharge of 28.8 mcm. Though the total storage of groundwater in the region is apparently high but in this area less mineralised water generally is encountered as lenses over more mineralised water. As such a major proportion of water in the storage is more mineralised water. The recoverable less mineralised water only comes from fresh annual recharge and little from an upper layer of stored water. Any attempt to overdraw water from the storage will lead to salt water intrusion. The water consumption by various sources for past ten years have been shown below:

(in mcm.)

	1966	*1974*	*1976*
Water requirement for human and livestock	2.651	1.640	2.494
Irrigation	4.816	5.920	9.543
Total	4.467	7.560	12.037

The drop in water consumption by human and livestock for the year 1974 is due to the heavy mortality and migration of livestock, which is approximately 1,07,166 cattle heads during the drought year 1968-69.

Assessment of Sensitivity of Desertification

Due to Water Exploitation

For evaluating any terrain in relation to its groundwater potential, three parameters are of paramount importance:

- yield from the wells
- salinity in the water
- depth of static water level below ground level

These parameters were observed during 1966 and again in 1976.To study the variation of these parameters and their interplay in relation to the terrain characteristics, statistical methods of weight point system has been adopted and the following scores were allotted to various parameters:

	Draft	Score	Salinity	Score	Static water level	Score	Surface water	Score
(a)	Wells stand draft for 4 hrs.	1	Over 6000 mmhos/cm	1	Deep-More than 12 m	1	Water in nadis last less than 6 months	1
(b)	Wells stand for 4-12 hrs.	2	2000-6000 mmhos/cm	2	Medium-6-12 m	2	Water in nadis last 6-9 months	2
(c)	Wells can stand draft over 12 hrs.	3	Below 2000 mmhos/cm	3	Shallow-Below 6 m	3	Water in nadis last 10-12 months	3

The various parameters of case study area had been classified on the undermentioned scale and its distribution studied:

Sensitivity	Score	Percentage area covered
Last sensitivity	12-9	10.67
Moderately sensitive	8-6	68.12
Highly sensitive	5-4	20.76
Hills	0	0.45

Effect of Induced Surface Instability on Soil Fertility

Increased human pressure on land through cultivation and overgrasing has manifested itself in instability of soil surface and acceleration of wind erosion hazard. It is of interest to know how this process affects soil fertility. Ideally, such a study can be made by comparing fertility status of soils in their pristine state and those under prolonged arable landuse. In absence of availability of naturally caintained habitat, other criteria were locked for namely: fertility status of sites showing differential manifestation of wind erosion hazard and fertility content of fenceline depositions in comparison to field proper. In course of this study 133 soil samples were collected and analysed for available phosphorus, potassium and organic carbon. Salient findings of the results are shown in the following table:

Name of soil mapping unit	No. of samples	Values of available plant nutrient elements			
		Organic	C%	Av P kg/ ha	Av K kg/ ha
Light textured soils of flat aggraded plains with moderate incidence of wind erosion	42	Mean	0.16	13.00	138
		S.D.		2.43	62
Light textured soils of hummocky aggraded plains with severe/incidence of wind erosion	27	Mean	0.14	10.76	134
		S.D.		4.63	68
Dunes with severe/incidence of wind erosion	18	Mean	0.11	10.72	128
		S.D.	0.06	4.96	68

S.D. = Standard Deviation

Results in the above table show that in comparison to moderately eroded sites, the hummocky and duny lands with a mass of loose sand and severe erosion have appreciable lower value of nutrient elements, particularly organic carbon and available phosphorus. In this study, soil samples were also collected from field proper and from soil piled along fenceline depositions had large intersite variation but in general their content of phosphorus and potassium was higher than the field proper (table below) indicating thereby that the latter is losing a more productive part in course of wind erosion. Organic carbon in fenceline deposition is less which probably is due to the fact that this fraction is not trapped by the fences.

Available Nutrients Element Content of Field Proper and Fenceline Depositions of Moderately Eroded Light Textured Soils

Site	No. of samples	Mean and standard deviation of available form of nutrient element			
		Organic	C%	Av P kg/ ha	Av K kg/ ha
Field proper	42	Mean	0.15	13.00	138
		S.D.	0.05	2.43	62
Fenceline	46	Mean	0.10	16.7	176
Deposition		S.D.	0.06	7.6	73

Both the results thus tend to show that increased incidence of wind erosion is leading to some loss of soil fertility though some eritical work in the field remains to be done. The above results perhaps find a confirmation in comparison of available nutrient status (table below) of soils under conventional farming system and conservation oriented grass cover. Sites under latter have appreciably higher content particularly of organic carbon.

Comparative Fertility Status of Ideally Managed and Conventionally Managed Sites

Site	No. of samples	Mean and standard deviation of available form of nutrient element			
		Organic	C%	Av P kg/ ha	Av K kg/ ha
Ideally managed	28	Mean	0.17	17.78	191
		S.D.	0.03	2.04	53
Conventionally managed	14	Mean	0.14	9.32	174
		S.D.	0.07	4.53	72

Ecological Changes in Vegetation

The desertification processes, by way of ecological changes, were studied in detail in Luni development block and the salient results are presented below. They indicate the presence of the recognised ecosystems, their susceptibility towards the deterioration, extent of degradation by projecting species composition and biomass production

Hills

A community of *Acacia senegal—Anogeissum pendula* dominates the hilly terrain kept under protection for a long period of time. But severe biotic factors like cutting, grazing, lopping etc. lead to a series of changes like severe water erosion, removal of the built up soil and exposing the hard rocks free of any soil cover. Degraded communities are represented by *Euphorbia caducifolia, Capparias decidua* community, and *Grewia tenax* community etc.

The community of *Capparis decidua* had 5.7 per cent cover 57.6 per cent relative dominance. Total plant density was 132/ha, which contributed 16.0 q/ha above ground biomass. A disturbed community of *Acacia senegal* with 3.5 per cent crown cover, 57.5 per cent relative dominance producing only 24.4 q/ha A.G.B. with 87 plants/ha density. *Ex-caducifolia* community had 132 plants/ha width.

A protected hilly tract of Kailana with *Acacia senega-Grewia tenax* community had 216 *Acacia* plants/ha with 47.5 q/ha above ground biomass as against 9.9 q/ha in unprotected areas, showing a net loss of 79.1 per cent.

Grassland of hilly regions are represented by *Tephrosia purpurea-Oropetium thomeaum* with 0.2 to 0.7 per cent ground cover and 0.35 q/ha above ground biomass, while protected area had *Eleusine compressa. Ariatida funiculata* community with 2.3 per cent cover and 15 q/ha forage production/ha.

Sand Dunes

Religiously protected dunes in the block are dominated by *Acacia senegul—Maytenus emarqinata* community which has 14.9 per cent crown, 40.4 q/ha above ground biomass and a total plant density of 228/ha. Slightly degraded community of *Maytanus* had 42.2 q/ha with 310 plants/ha and 11.0 per cent crown cover. A degraded community of *Acacia Jaccuemontii Lycium Barbarum* on stabilised dune had 282 plants/ha but with 5.7 c/ha above ground biomass. Another degraded shrub community of stabilised dune—*C. deciduas, Balanites aeqyptiaca* showed 376 plants/ha with 3.7 q/ha above ground biomass. Therefore, the data shows a reduction of about 9.13 per cent in biomass production and 26.8 per cent shrinkage deficient of perennial grass. Annual like *Cenchrus biflorus* contributed 7 to 11 q/ha whereas the protected grasslands on sand dunes contribute 43.0 q/ha forage. Absence of perennial grass species results in quick sand movement during summer months and its movement leads to quicker desertification process.

Piedmont Plains

A well developed community of *Saladora oleoides, Prosopis cineraria, Capparia decidue* had 213 plants/ha with 10.3 per cent crown cover and 3.35 q/ha above ground biomass whereas degraded community of *Capparia decidua* alone had 4.3 per cent crown cover, 197 plants/ha and 24.5 q/ha. The grassland deterioration was depicted by *Oropetium thomeaum, T. purpurea* with 0.5 to 1.0 per cent ground cover and negligible above ground biomass 0.14 q/ha.

Pediment Plains

A seral community of *E. caducifolia-S. oleoides* under partial protection had 9.5 per cent crown cover with 63.7 q/ha above ground biomass whereas in disturbed conditions, the same area was represented by *Capparis decidua* bushes as most of the rocky parts were completely exposed. The grassland status was similar to piedmont plains.

Flat Older Alluvial Plains with Light Soils

Large area of the block is occupied by these plains which are largely put for cultivation year after year. Cultural practice permit only few trees of *Prosopis* in the field. A density of 20-48 plants/ha showed 14.0-45.3/q/ha above ground biomass of trees. Few 2-3 years fallow fields showed a high regeneration of *Prosopis* 54-258 plants/ha indicating that a very good forest can develop in the absence of cultivation. Some of the areas left as grazing lands are invariably dominated by *Capparia decidua* with scarce presence of *prosopis* trees. One well protected community had 168.1q/ha above ground biomass with 254 per cent crown cover and 282 plants/ha population. Similar community in body degraded condition supported only 2.8 q/ha above ground biomass, 5.7 per cent crown over and 232 plant/ha. Thus, biomass and 77.6 per cent shrinkage in crown cover. Desertification process is very much evident from these figures. Ground flora is mostly represented by *Tephrosia and C. biflorus*

Flat Older Alluvial Plains with Medium to Heavy Textured Soils

Here predominant plant community is of *Prosopis cineraria-Salvadoru oleoides* which forms the climate and climate of plants but nowhere does it represent its optimum expression. The plant density ranges from 12 to 20/ha whereas *C. decidua*, as chief associate biomass production ranges from 63.8 to 71.3 q/ha with 5.5 to 7.0 per cent crown cover. In a semi-protected area, the biomass was as high as 146 q/ha with 15.3 per cent crown cover having 91 plants/ha only. Excessive grazing leaves *Tephrosis purpurea* should dominate whereas *Dichanthium* and *C. setigerus* should express themselves as dominating grassland community.

Sandy Undulating Older Alluvial Plains

Similar to flat plains *Prosopis-Capparis* community largely dominates but with lesser plant density (10-15 plants/ha). The values of above ground biomass vary from 7.6 to 14.4 q/ha with 19 to 32 plants/ha. Grazing lands with *C. decidua* as the dominating community support 55.1 q/ha above ground biomass with 302 plants/ha. A similar degraded community with 232/ha yielded only 16.7 q/ha above ground biomass. *Crotalaria burhia* and *Aerva pseudotomentosa* predominates fields whereas *Ariatids* and C. *biflorus* dominate the cultivated area (1-1.6% cover). *Lasiurus sindicus* should have been the climax grassland of this ecosystem.

Younger Alluvial Plains

These are confined along the course of river Luni and its tributaries. *Salvadora oleoides, Prosopis cineraria, and C. decidua* dominate this ecosystem. *S. oleoides* had 75 per cent relative dominance and 21.1 to 31.1 q/ha above ground biomass. Here also the vegetation remains in degraded condition due to continuous cultivation. *Acacia nilotica*, which was quite dominant at certain spots before 1939 are now very much degraded and without any tree of *Acacia desmostachva* forms main grassland community with 5 to 10 per cent crown cover.

Saline depressions formed due to chocking of small streams outlet, are mostly devoid of any vegetation. *Soorabous marginatus* forms the dominant grassland community whereas *Capparis decidua* either grow on hummocky or on the outskirt depression. In patches the grass *Soorobolus* had 12.8 per cent ground cover and

11.0 q/ha above ground biomass during 1967 but now (1976) showed only 4.7 per cent ground cover and 5.7 q/ha above ground biomass. Present enquiry revealed that the severe famine of 1969 adversely affected this grass which could revive to this extent only. Here the desertification process is due to droughts, heavy grazing and salinity.

From the foregoing observations, it is very obvious that the desertification processes have been triggered off and conditions are deteriorating continuously leading to high degree of ecological imbalance and accentuated wind and water erosion.

Landuse Changes

The above figures indicate that there are no significant changes in areas in different landuse units. The processes of desertification remain unchanged and the man-induced processes (effect of the urban centre of Jodhpur) have not produced any significant results in negative desertification in this block. Rather the increased cultivation in the marginal lands and uncontrolled grazing in the unfenced permanent pasture or oran lands have further deteriorated the desertic conditions in many places, particularly in the northern track. The use of brackish water along the riverine tracks of the Luni and the Jojri river for irrigation to raise crops have further affected the desertification processes as the land under saline wastes has been considerably increased in the recent years particularly in Lunawas area and along the bank of the Luni and the Jojri rivers. In the northern part fresh depositon of sands along the foothills are also evident in Rohida Kalan areas.

Intensity of Desertification

On the basis of 13 maps prepared by the different sections of Basic Resources Studies Division and data obtained from climatology and sociology sections, a desertification hazards and processes involved in desertification in different areas of Luni Block has been cartographed. It has been assessed that 20.11 per cent of the total area of the block constitutes arid zone. The entire arid zone and a greater portion of the semi-arid zone is affected by desertification hazards. About 971.50 sq. km or 46.12 per cent of the total area is high to medium vulnerable due to enlargement of salted areas, dunes, sandy cover, shifting sand and hummocks. Rest of the areas are affected mostly by water erosion hazards associated with dunes or salinity or both. 53.88 per cent of the total area of the block is affected by medium to slight vulnerability. Dune is the most important process of desertification affecting 39.1 per cent of the total area. In the rest of the area, it is associated with either water erosion (1.6%) or poor germination of seeds (7.4% or both 2.6%).

Arid zone is affected by relatively very high density of population. The density of both livestock and population are relatively high in the semi-arid zone. But the latter bring the area under medium to slight vulnerable lands—mostly constituting medium desertification hazard areas. Desertification hazards are very high in the north-west part and in an elongated belt in the south-west of the semi-arid zone. Very high desertification hazards area constitutes 590 sq. km or 29.66 per cent of the total area. 2.13 per cent of the total area is demarcated as slightly desertification hazards belt for which either the figures of livestock and human population are not available or where their relative pressures are very low. These areas (2.13%) are also marked by slight vulnerability. It has been estimated that 223.80 sq. km or 11.30 per cent of the total area can be controlled easily—control measures by groundwater irrigation are already under process. But here too, recharge methods should be adopted for long-term landuse management as water resources are depleting and salinity is increasing as a result.

Need for Diagnosing and Monitoring Desertification

It is clear from the foregoing account that there is an alarming decline in the resource base through overuse by both human and livestock pressure and if unchecked a stage of entropy will be reached which is irreversible. The result would be a catastrophic crumbling of the resource base and the environment will eventually become unfit for man's occupancy. For these reasons, desertification must be diagnosed and monitored for the whole range of physical, biological and social processes relevant to the ecosystem. The human factor is a central consideration in these studies.

Desertification Processes to be Monitored and their Impact

Desertification is discontinuous in time and space. It is necessary to take into account the seasonal variations (ephemeral or annual plants), annual variations (e.g., yield of crops), great climatic cycle (drought period) and abrupt variations (change in landuse). To foresee the trends in desertification usually a long set of observations is necessary.

Desertification is also discontinuous in space (e.g., desertification around a well, or around a town in an overpopulated region). The phenomenon will be more pronounced in the fringe areas of the desert adjoining the semi-arid region where ecotonal gradient is at maximum.

Process of Desertification Monitoring

The following are the parameters indicative of the various features to be observed in a time sequence in the field and on aerial photo or satellite imagery:

1. Moving of dunes.
2. Deflation of fine material.
3. Sand cover and beginning of sand accumulation.
4. Enlargement of rock outcrops or indurated horizon on surface caused by erosion.
5. Enlargement of bare areas because of sealing of soil surface.
6. Salinisation.
7. Water erosion.

Impact of the Processes

1. Degradation of vegetation.
2. Yields of crops of pasture decreasing.
3. Changes in water balance.
4. Frequency of dust storms.

Factors Causing the Processes

1. Climatic changes.
2. Human and animal pressures.

Approach to Undertake Monitoring Activities

The approach to monitoring the phenomena on the earth's surface usually requires there sequential types of activities namely: (i) the identification and location of the areas to be monitored, (ii) the survey of present condition in those areas, and finally (iii) the repeated observations of those areas which constitute the monitoring proper.

Identification and Location of the Areas

In order not to overlook current knowledge and activities, it will be necessary to consider the available information mainly in the literature, but also available from appropriate research institutes and other relevant sources.

Using maps and documents on climate, vegetation, soils etc., as well as remote sensing data, the first step will be to trace the boundaries of arid and semi-arid zones in the region under consideration.

Having determined and mapped the boundaries of the macro-areas, it may greatly simplify subsequent steps if these boundaries are transferred to satellite imagery in the form of a mosaic of the entire area, which essentially constitutes a photo-map. Such photo mosaics can be prepared relatively cheaply from Landsat (either band 5 or 7) black and white imagery in convenient scale of 1:100,000 or 1:250,000.

To facilitate selection for monitoring, the areas thus delineated could well be sub-divided on the basis of various sources of information into classes of desertification and degree of socio-economic impact. Thus, an area classified as having a high degree of active desertification with a high socio-economic impact would be clearly self selected for monitoring whereas an area with a moderate degree of desertification and of low socio-economic impact would have a much lower priority.

Survey of Present Conditions and Provisions of a Data Base for Monitoring

Planning for monitoring involves a survey of present conditions which must include consideration of the processes which are currently leading to desertification. Among these may be climatic changes, soil erosion, depletion of water table, removal or degradation of vegetation by grazing and/or cultivation pressure or poor land management practices.

Having determined the causes of desertification either from existing information or by low intensity multi-stage sampling on the ground and aided by aerial photographs and satellite imageries, the factors to be served in order to monitor desertification process or their onset can be identified.

Surveys are envisaged at map scales between 1:150,000 to 1:250,000 depending upon various factors including the nature and size of the areas selected and the need to execute remedial or preventive measures in view of the socio-economic implications. In places where good topographic maps are not available, enlarged landsat imagery possibly annotated with some planimetric detail not readily visible, may be used as a base map. In any case, up to date base map at selected scale will be required for the initial surveys.

Available and required information must influence the determination of the scale at which the survey of each significant factor should be carried out.

In addition, there are likely to be factors such as the level of groundwater and population growth (both human and livestock) for which statistical recording by geographical coordinate will be more appropriate than mapping alone. This, in turn, requires consideration of the desirability of establishing a data bank to store, manipulate and retrieve the survey and subsequently monitoring data, a decision which should preferably be taken at the outset since it will affect the presentation of data to be collected.

Another implication of the scale of survey for different factors may use satellite imagery in association with air photos and ground sampling. Detailed land use may be mapped from small scale high quality aerial photos (colour infrared diapositives), while yet other factor like groundwater levels can only be surveyed on the ground.

Since all the considerations discussed above are clearly interrelated, these should be approached holistically when planning the survey of present conditions.

Monitoring

Essentially monitoring will consist of repeated surveys of the type undertaken to determine the present condition and, therefore, it will be necessary to determine the interval of such repeated surveys. Very short or very long intervals are likely to be irrelevant, e.g., daily weather pattern or major changes in land form. Relevant interval for periodic monitoring will be determined by the rate at which various elements of the desertification process change to a significant and detectable degree.

Earth observation satellites presently provide cover at nine to 18 days interval for areas by a regional receiving station. Another monitoring facility is provided by the capability of the earth observation satellite to act as data collecting platforms by receiving transmissions from automatic recording stations on the ground and relaying those to a central data collection facility.

The capacity for monitoring a range of interval is, therefore, considerable and able to accommodate the needs likely to emerge as a function of the rates of changes to be observed. To determine the optimum monitoring procedure is evidently a complex matter, best approached by making relatively frequent and large scale observations and progressively refining these on the basis of experience and of change in the factors being monitored.

Ecological Principles in Combating Desertification

Having considered thus far the evidences, processes of desertification, their impact and monitoring thereof, it is perhaps logical to consider some ecological principles in combating desertification.

Conservation in Dry Ecosystem

Conservation being a policy of sustained yield, it should be the basis for action in combating desertification. To this end, it is necessary to keep in mind the following ecological principles:

(1) Comprehensive planning is necessary to maintain the all-important integrity of the ecosystem. Traditional landuse strategies are often, but not always integrated with the ecosystem and their change may endanger even quite remote facets of the system.

(2) Where investment is meagre, it is better to cooperate with the ecosystem than to fight it. Where investment is available, higher yields can be achieved by making modifications to the ecosystem but new strategies must be evolved to counter degradation.

(3) Long-term sustained yield is the only sensible management of dry ecosystem even if considerable fluctuations from the mean are inevitable. Resting and fallowing are important tactics where output is low and unpredictable.

(4) The economics of dry land cannot exist on its own but given fair processes of change can contribute useful produce to

the wider economic system. The most effective use of dry lands depends on the flexible use of many complementary facets.

(5) Land capability assessment is a basic tenet of any conservation policy but such a policy should be both dynamic and flexible.

These principles apply very well to the pastoral ecosystem and to a lesser extent dryland ecosystem. In a highly capitalised system such as irrigated agriculture where the ecosystem is vastly altered and bolstered by technology these ecological principles have little real meaning.

Irrigated Agriculture

The pattern of desertification followed in many irrigation schemes includes the Indus valley irrigation scheme in Pakistan. It is usually true that only when the problem of water logging and salinity suddenly appear on a wide scale and when production is threatened, that programmes of drainage are begun. Once the canals are installed, the digging of drain or tube wells should be immediately started in anticipation of the problems that will inevitably occur, with of course suitable planning for terrain characteristics. The pay off would come quite quickly.

4

Integrative Mechanism in the Desert Development Administration in Rajasthan: A View from 1980

RAMESH K. ARORA

Rajasthan has a vast desert area covering about five lakh sq. kms. and encompassing the districts of Bikaner, Jaisalmer, Barmer, Jodhpur, Nagaur, Churu, Pali, Jalore, Ganganagar, Jhunjhunu and Sikar. This ecological constraint has put a marked imprint on the quality of life of the people in this area. Poverty is the inevitable upshot of inadequate rainfall, recurrent famines, constant migration of men and livestock, unfavourable soil conditions and a host of related factors.

The integrated approach to desert development adopted in Rajasthan, as in other states, includes a battery of measures such as afforestation, soil conservation, animal husbandry, dairying, irrigation schemes and rural industrialisation. Such a comprehensive methodology for the development of desert areas has necessitated a strategy of planned change with a concentration on an integrated administrative mechanism that can effectively facilitate such a change. This note focuses upon an analysis of the adequacy of the integrative elements in the present administrative system for desert development at the state and district levels in Rajasthan. At the outset, a brief reference to the existing administrative pattern may appear to be pertinent.

Existing Set-up

A unique feature of the administrative system in Rajasthan is the functioning of the Special Schemes Organisation at the state level which was set up in 1973 with a multidisciplinary character and having responsibility for undertaking and directing project planning and administration pertaining to certain special schemes. Over the years, the role of Special Schemes Organisation (SSO) has substantially increased; its multi-functionality should be evident from the fact that it has now assumed the responsibility of implementing a large number of massive sized developmental schemes such as Integrated Rural Development, Antyodaya, Samagra Gramin Vikas, rural growth centres, Desert Development Programme (DDP) and Drought-Prone Area Programme (DPAP).

It must be stressed at this point that the Special Schemes Organisation is an important administrative innovation in the sense that a developmental organisation of this magnitude which has been set up outside the traditional departmental arrangement, provides it with considerable flexibility in its structural designing, personnel administration and functional operations. The multidisciplinary specialists are recruited in the Special Schemes Organisation on the basis of special selection rules for a fixed tenure. These specialists, on completion of their tenures, are expected to go back to their parent organisations with an enriched background and experience in project formulation, management and monitoring. Briefly, the Special Schemes Organisation formulates projects for three categories of schemes :

(i) schemes designed to increase production and enhance productivity in agriculture and allied sectors;

(ii) schemes relating to area development; and

(iii) schemes for the amelioration of the status of the rural poor.

The schemes for desert development and drought-prone area programme fall in the second category.

The Special Schemes Organisation is headed by the Secretary, Special Schemes who has an elaborate staff under him consisting of deputy secretaries, project directors, officers on special duty, project officers, project economists and other functional area specialists.

Out of four deputy secretaries, one deals with drought-prone area programme and project formulation while another concerns himself solely with desert development programmes.

Besides the Special Schemes Organisation, which is the most prominent agency for the planning, management and monitoring of desert development programmes, another state level agency involved in the desert development is the Desert Development Commissioner, the headquarters of which is situated at Jodhpur. The Desert Development Commissioner having the status of head of department, is supposed to supervise, coordinate and monitor the desert development programmes in all the desert districts, although its power and authority have generally remained at a low key.

The major focus of project implementation, however, shifts to the district level where district rural development agencies (DRDAs), registered under the Registration of Societies Act, operate. These agencies, set up in all the districts of the state, are headed by the District Collector with MLAs of the district concerned, district level officers of the functional departments, and heads of the financing institutions as their members. The governing council of a DRDA is expected to meet periodically, at least twice a year, to review the progress made in various developmental schemes assigned to its area of operation. The district rural development agency is the basic planning unit and is responsible for implementation of special schemes. It assigns specific tasks and targets to various departments and institutions and releases funds to them.

Monitoring of projects is an important responsibility of this agency. Notably, all the state funds for the implementation of special development schemes, including desert development and drought-prone area programmes, are placed at the disposal of this organisation. An advantage of this approach is that it obviates lapse of funds at the end of the financial year and therefore, continuity in financial planning becomes much more feasible. More important, however, is the fact that a DRDA enjoys much more financial autonomy than a government department operating through its district level officers. Thus, the involvement of various departments in project implementation is obtained largely through their liasion with DRDA. Besides, the departments place officers of their cadres in various positions related to their respective functional areas at the district level. This way, departments have a major say in selecting the type of personnel that man the various projects at the district level.

It may be mentioned that each DRDA has a project director who works under the overall supervision of the Collector. The post of the project director is a crucial position, since the basic function of project formulation and evaluation is entrusted to this officer who is expected to coordinate a number of special schemes that are executed simultaneously in one single district.

The Project Directors working in desert districts are recruited through a committee formed particularly for this purpose. Five districts have IAS officers as project directors, while the rest of ten districts have Project Directors of RAS cadre. Besides the project directors, the subject matter specialists, assistant project officers, district planning officers, project economists, credit planning officers and a few other officers who are selected through an open selection by the concerned technical departments are posted to DRDAs.

Problem Areas in Integration

The desert development administrative system is beset with a few noticeable limitations in the realms of control, coordination and responsibility. The maxim that authority and responsibility should be mutually commensurate in an organisation appears to have been grossly under-emphasised in this crucial area of state administration.

The Desert Development Commissioner, for instance, is supposed to supervise, coordinate and monitor desert development schemes in all the desert districts but he has no administrative or financial powers for enforcing discipline in the implementation of various projects. It is only the status and position supertime IAS of the Desert Development Commissioner that lends him informal authority which he may use in persuading the chairmen of DRDAs to act according to given scheme of things and in consonance with the state policy. The Desert Development Commissioner is devoid of essential formal authority of supervision and control over project implementation in DDP. Of course, his reports to Special Schemes Organisation will carry weight as far as his assessment based on monitoring is considered.

To a great extent, the Secretary, Special Schemes Organisation is also the head of the department of his organisation and, therefore, he has to perform major supervisory responsibility in this context. It is the Special Schemes Organisation which releases funds to various functional heads of departments for the creation and continuation of various positions of departmental functionaries at the district level for the implementation of special schemes. In addition, it is the organisation which assigns various projects to DRDAs. But a complex situation arises when SSO can exercise no control over various heads of departments regarding the type and appropriateness of the personnel to be placed at the district level. Further, it has been observed that on many a post, the required functionaries are not appointed on time and the posts remain vacant for considerable duration to the detriment of project effectiveness. Not surprisingly, therefore, shortage of staff has been a major bottleneck in the efficient implementation of development projects. And with quite a few other missing links in personnel administration, the whole administrative system at the district level suffers. Interestingly, the DRDAs too cannot be held responsible for any ineffective implementation of projects, for they too have no controlling authority over the whole area of personnel placement. Even for a large number of posts within the DRDAs, the choice of personnel rests with various departments and it has been observed many a time that less-committed and less-experienced persons are posted in the field; and to cap it all, a large number of these officials are found continually engaged in the profession of managing their transfers.

An irritating factor at this level of personnel administration relates to the absence of any codified service rules for non-gazetted and ministerial staff of DRDAs. As a result, many persons with lesser duration of service get confirmed earlier in various government departments than their counterparts in DRDAs. This causes demoralisation in the rank and file of DRDAs and thus makes them less attractive places for taking up jobs.

An important issue area in the implementation of various special schemes (including DDP and DPAP) is that the heads of departments, who are assigned tasks by DRDAs, in the absence of sufficient delegation of authority, have to seek sanctions for spending on projects from their respective secretarial departments and which, in turn, have to seek the approval of finance department.

This whole process is dilatory and to a great extent redundant. Moreover, delays incurred in making necessary purchase has an accelerating impact and adversely affect the whole time-schedule of project implementation. Even more serious is the point that departments/corporations/boards which are entrusted with the task of implementation of projects are given very little role in preparing plans, deciding about the projects location and in transfer of funds from one sector to another.

The extent of centralisation of authority is also indicative from the fact that in the process of approval of plans, programmes, projects and priorities, very little importance is given to the projects submitted by DRDAs and most of the decisions regarding the initiation of projects are made without taking recourse to a deeper analysis of regional problems and the specific requirements of areas as perceived by DRDAs and their senior project staff. Not only does this system hamper the process of involvement of district agencies in the overall project planning but it sometimes takes the projects farther away from realism. It also looks paradoxical that an organisation such as DRDA which is given almost no role in planning process is held accountable for the implementation of projects.

Further, it has been observed that no rigorous exercises are undertaken at the initial stage of the project to identify clearly the priorities and problems involved in the whole exercise of development administration and as a result a premium is put on ad-hocism and improvisation in the area. Hasty planning is faulty planning and needs to be replaced by meticulous and detailed project planning.

Ironically, the DPAP has been treated not as a comprehensive, well-integrated are a development programme but only as an adjunct to the total developmental planning. Devoid of an intensive and profound need and resource-based planning effort, it provides at the best a partial view of the canvas of development and cannot be termed as an 'integrated' planned effort in the true sense of the term.

It may be pertinent to mention at this point that the multiplicity of agencies, viz., SSO, desert development commissioner, heads of departments, staff of directorates and district rural development agencies, has created a diffusion of inter-level authority-responsibility relationship and this has resulted into a low level of effectiveness in project administration in DDP, DPAP and other realms. Although, it may not be feasible to reduce the number of agencies involved in the operation of DDP and DPAP, it would seen desirable and practical to bring about greater specificity in inter-role linkages in the total special schemes developmental enterprise. Specifically in matters of desert development programmes, there is a visible overlapping and duplication of monitoring of project implementation. On the one hand, the Desert Development Commissioner engages himself and his staff in this task, while on the other hand, the SSO also takes serious initiative in monitoring the projects through its senior officers. This dual monitoring on most occasions, can become wasteful.

The state of diffusion and role unspecificity has also resulted into many lacunae in the total control systems. For instance, there is no uniformity in the accounting system of various DRDAs. Besides, this aspect of financial administration is rarely checked by any competent authority. In addition, there seems to be no specified system of internal checking or auditing of the financial operations of DRDAs.

Possibilities of Redesigning and Adaptation

In the following lines, certain broad suggestions of a tentative nature are being given only with a view to provide an opportunity in fresh thinking about the possible areas where administrative adaptation may be possible. Most of these suggestions are a priori in their character and need to be supported (or modified) further on the basis of rigorous empirical research which does not appear to have been undertaken so far in this sphere. All that is being stated below is not entirely new; certain points have been restated only with a view to stress their relevance to the broader objectives of making DDP more effective in its upshot.

Moreover, it may be hinted that no organisational redesigning scheme is being proposed here primarily because such schemes have meaning only when an advance holistic and systemic analysis has been made in the context of the entire politico-economic and administrative setting. And secondly, such an ambitious schemes of change is generally underrated and considered as too 'academic' to fit in the 'existing' and time-honoured framework of administrative apparatus.

1. The whole process of project planning, implementation and monitoring needs to be provided a reinforced leadership by Special Schemes Organisation. The secretary, SSO should involve desert development commissioner, heads of departments, collectors (chairmen, DRDAs), project directors of DRDAs and its own specialists in periodic meetings to facilitate rigorous planning and effective coordination. Presently, there is greater stress on meeting the heads of departments than the desert development commissioner and DRDAs, and this imbalance needs to be corrected.

2. It is only after detailed discussions and deliberations at the level of committees referred to above that priority sectors should be specified and on their basis, detailed plans should be evolved. Such plans then should be placed for approval before the respective sanctioning committees.

3. Plans should be transferred to DRDAs for implementation soon after these have been sanctioned so that DRDAs start taking initial organisational steps to implement them. It may be stressed that district level projectisation needs to be taken up with a holistic perspective and with profundity. DRDAs should not concern themselves only with monitoring. Their role in total project administration should be effectively strengthened. At this crucial stage, the desert development commissioner should also be involved in a formal or non-formal manner.

4. DRDAs should arrange with the heads of departments for the execution of various approved projects and provide them necessary financial resources. In this process, the desert development commissioner may be actively involved. Notably, he can act as friend, philosopher and guide to the chairman of DRDA because of his overall 'macro vision' and by virtue of his close liaison with, SSO which needs to be further cultivated.

5. Sanctions should be issued by SSO in the name of DRDAs for various works and in the name of heads of departments for various staff positions.

6. Certain specific small-scale and medium-scale works can be undertaken by DRDAs themselves. The selection of works will depend upon a particular DRDA's capacity. For such projects, sanction for works can be issued in the name of DRDAs themselves.

7. DRDAs should continually monitor the implementation of projects in association with the desert development commissionner. In fact, he can act in complete unison with SSO in matters of monitoring and thus help obviate overlapping in the monitoring function. While the chairman of DRDA should engage in monthly monitoring, the desert development commissioner should undertake this exercise at least every quarter. The SSO's project monitoring staff should also be strengthened to support the total monitoring system.

8. The secretary, SSO should conduct quarterly the meetings of coordination committee with desert development commissioner, DRDAs and heads of departments, preferably just before the meetings of the State Planning and Development Coordination Committee.

9. Greater delegation of financial powers should be made to heads of departments to minimise references made to the secretariat. Likewise, as far as possible, only cases pertaining to creation of staff positions should be sent to the secretariat for administrative and financial sanctions.

10. State government should approve the rate contract rates for purchases and let the actual purchases be made by DRDAs/ district level officers.

11. It should be left to DRDAs to collect utilisation certificates from the various executing agencies, get the accounts audited by chartered accountants and transmit such audit reports to the state government. However, the system of accounting and auditing of DRDAs should be systematised and regularised. Besides, periodic inspection of the records and accounts of the DRDAs by the Registrar of Societies should also be undertaken. This will help tighten the whole administrative apparatus.

12. Posts for implementation of various programmes and projects should be created in concerned departments in consultation with DRDAs and DDC. The head of department controlling specific cadres should be sounded for making advance planning concerning the filling up of various posts created for the implementation of special schemes projects. In placement of officers in districts as well, SSO and DDC may keep liaison with the heads of departments and accord particular importance

to the experience, aptitude and organisational commitment of officers. It may not be considered parochial to suggest here that there should be a regional bias in the selection of officers in districts. This should also apply to the recruitment to various middle level and junior positions. A regional bias in recruitment will also mitigate the chances of early transfers and quick turnover. It may be difficult to codify this clause in a formal manner but it can certainly be kept in mind while making selection for various district positions.

13. Some rules for various non-gazetted and ministerial staff of DRDAs should be finalised so that confirmation of the staff may not get unduly delayed.

These are only a few points for strengthening of integrative mechanism in the desert development administration in Rajasthan. The catalogue of points is by no means exhaustive nor has it been drawn on the basis of a rigorous detailed empirical investigation. However, these do indicate the possible areas of action in the sphere of tightly integrating a hierarchy of agencies engaged in desert development programme.

5

The Developmental Delivery System in a Desert District in the 1980s

RAKESH HOOJA

Much thought is being given as of the early 1980s[1] towards improving the governmental rural development delivery system in India. However, except for discussions of specialised project agencies like an SFDA (Small Farmers Development Agency), or a DDA (District Development Agency), or of the developmental role of a District Collector,[2] or of the earlier role of Panchayati Raj when it was virile,[3] this author has not come across an attempt to take a holistic view of the developmental delivery system in a Desert District. Hence it is proposed to describe not only the role of the DDA (now rechristened District Rural Development Agency – DRDA) in implementing the Government of India special programmes of DDP (Desert Development Programme) and IRD (Integrated Rural Development) which now incorporates the TRYSEM (or Training of Rural Youth for Self Employment) Scheme and the state government programmes of Samagra Gram Vikas (whole village development), Antyodaya and rural growth centres; but also how famine works are administered in a district and how far the Minimum Needs Programme and other "normal" works of various government departments are coordinated at the district level. Linkages with state level administrative units shall also be discussed.

II

However, before commenting in detail on the functioning of rural development administration at the district level, it would be proper to set forth a general list of the various constituents of the rural delivery system and to categorise some government programmes as area-based or target group-based as also to link specific programmes with the one or many of the constituents of the delivery system that administer the programmes.

The following are the constituents:

(1) Revenue Agency

This is a popular term for the land revenue administration based system of the District Administration headed by the District Collector. This is normally saddled with the responsibility for execution or coordination of all sorts of government schemes as also with all aspects of regulatory administration.

(2) Departmental Staff

This in a way is a part of the above category since staff belonging to the various departments of government (including corporations and boards) i.e. irrigation, agriculture, animal husbandry and the like, are coordinated at the district level by the Collector in addition to being controlled by their state level heads of departments.

(3) Development Agency

This is also a popular term used in the field and includes two sub-categories :

(a) Panchayat Samiti officials, and

(b) the elected panchayat non-officials

Both are inter-related and have close links with the revenue agency as the Collector is an important part of this sub-system also. At present with Zila Parishads and Panchayat Samitis superceded in Rajasthan, the elected non-officials exist only at the level of the Gram Panchayats.

(4) New Agriculture Extension Agency

This has recently emerged as the village level extension staff has been detached from the development agency and set up independently under the state agriculture department.

(5) Special Projects

These have been specially set up by the state or Central government for special purposes and have to often rely on all the above listed agencies as also on the next category to a large extent. Collectors are normally associated with them (e.g., SFDA, DDA, CADA etc.) in some capacity or the other.[6]

(6) Institutional Credit System

This comprises of the cooperative and commercial banks as well as regional rural banks, whose attempts supplement those of all the above categories and who in turn draw from various refinancing agencies like ARDC, NCDC etc.[7]

(7) Voluntary Agency

These often use government funds to execute government approved programmes/projects.

(8) Beneficiary Associations

Government aided or sponsored water users associations, milk producers/rural artisans/Scheduled Castes/cooperatives etc. committees to monitor shortcomings in government's land development programmes are other examples.

(9) Normal Trade Channels

These are often used for supply of inputs etc.

(10) The Mass Media Network

Government programmes, benefits and delivery system constituents are presented in Table 1.

III

Having given a general idea of the delivery system it is appropriate now to turn to how it works with reference to the programme being implemented in a DDP district though it must be added that the structure is similar for a non-DDP district also. As the purpose is to comment on the prevailing system and not merely to describe processes, the description shall be simplistic in nature.

Table 1

Government Programmes, Benefits and Delivering System Constituents

Type of Government Programme	Type of Benefit	Constituent of Delivery System
1. Target Group Oriented (a) SFDA, IRD,[8] TRYSEM, Antyodaya[9] (based mainly on subsidies and institutional finance as well as extension demonstration and training) (b) Bonded Labour Rehabilitation, Scheduled Castes Development, etc.	*Individual benefits:* (i) Credit and subsidy for livestock, minor irrigation, inputs, land development, improved implements, rural artisans, (ii) training for self employment, (iii) old age and disability pensions, (iv) provisions of employment *Community benefits:* (i) Loan and subsidy for community irrigation, market, yard development, land development and soil conservation watershed works, (ii) Support to Amul type dairy schemes like managerial subsidies for milk producer's cooperatives.	Revenue Agency, Development Agency, Departmental Staff, New Agriculture Extension Agency, Special Projects Institutional Credit System.

Contd.

Contd.

Type of Government Programme	Type of Benefit	Constituent of Delivery System
2. Comprehensive and Integrated Area Oriented	*Individual benefits:*	Special projects with assistance of other constituents of the delivery system.
(a) (i) CAD-based schemes which also have links with SFDA, IRD. Execution of works and provision of credit and subsidy are part of the programme.	(i) assured and timely water supply as per rational norms, (ii) agriculture extension message, (iii) inputs including credit, (iv) drainage system including field drains, (v) farm roads, (vi) land development, (vii) Consolidation of holdings.	
	Community benefits:	
(ii) Tribal Sub-Plan Area Development Programmes (though at times these are implemented on an unintegrated sectorally differentiated basis). This also includes MADA for tribal mini blocks in non-tribal areas.	(i) Protection of table lands through soil conservation, land development and afforestation, (ii) market yard development, (iii) development of roads and new settlements in virgin areas receiving irrigation for the first time.	

Contd.

Contd.

Type of Government Programme	Type of Benefit	Constituent of Delivery System
(b) Samagra Gram Vikas and Rural Growth Centres.	*Community benefits:* (i) Provision of infrastructural facilities and of services. *Individual benefits:* Similar to those for IRD, SFDA, CAD, Antyodaya and use of services.	Whatever agency is found suitable and available, especially departmental staff.
3. Multi Sector (Though Sectorally Implemented Area Oriented) (a) DPAP/DDP (b) Food For Work (FFW) and Famine, Minimum Needs Programme (Food for Work is now being replaced by National Rural Employment Programme (NREP).	Same as in 2 (a) above	Coordinated by special project though chiefly implemented through departmental staff, except for IRD, SFDA type work. Chiefly departmental staff though in FFW and famine some role of Development Agency is there.
4. Departmental Works may be Area of Beneficiaries Oriented	Depending on department.	Departmental staff

Government planning[10] today chiefly consists of what may be called "accountants planning". As per indications of the Planning Commission the state planning department fixes sectoral outlay ceiling which are communicated to secretariat departments who, jointly with the state finance and planning departments, decide in Budget Finalisation Committees (BFCs) the items (committed and new) which are to be taken up with the available money. The secretariat or administrative department passes on the funds and targets to its executive departments who pass them down to divisional and district level officers for implementation and monitor physical and financial progress. The targets may or may not have been based on proposals from below.

Money for DDA, DPAP, IRD, SFDA etc. is passed on in Rajasthan to the agriculture (Special Schemes Organisation) department in the secretariat. This department does not have an executive department and functions as its own executive department. It, therefore, passes on its funds and expenditure and physical achievement targets either to the DRDA or directly to various other executive departments for use in particular areas for specified jobs. The DRDA also in turn administers some of the funds for IRD, SFDA, TRYSEM directly with the help of the development and revenue agencies and passes the rest on the various district level officers of the various concerned executive departments, who have also already received both DDP etc. funds from their own executive departments as well as the departmental funds for departmental schemes.

The system for Samagra Gram Vikas (SGV) and Rural Growth Centres (REC) is somewhat different as executive departments are expected to use some of their own departmental funds for SGV or RGC project work and the Special Schemes Organisation only passes on small amounts in the form of additionalities to the DRDA for implementation of SGV/RGC projects. For the Minimum Needs Programme the state plan department is the model department and it passes on funds to the concerned executive departments through their administrative departments for providing select services and facilities like roads, schools, hospitals, electricity, drinking water, and house sites in villages on the basis of population.

The state plan department monitors progress and coordinational problems of all plan programmes through quarterly State Level Planning and Coordination Committee (SCPCC) meetings presided over by the Chief Secretary which are held for each department/ group of departments. In additional the Collector as head of various district level committees has the task of coordination. The Collector being chairman of the DRDA also tries to coordinate all DRDA activities. The Collector's coordination depends to a large extent on personal factors as departmental district level officers are subordinate to executive departments and provisions for Collectors making comments in annual performance reports of district level officers (DLO) keep being added or deleted. Any other coordination has to be through specially called meetings at appropriate levels.

Whenever a famine is declared the state famine department which, like the Special Schemes Organisation, is both an administrative and an executive department passes funds and delegates administrative powers to Collectors (and for small works to Gram Panchayats) who get famine works done either through concerned district level officers or through the development agency with the revenue agency assisting in supervision. A district famine advisory committee also plays a major role.

The Desert Development Commissioner also reports to the Special Schemes Organisation. But, as he has no administrative powers over either the executive departments, or the Collectors and DRDAs, or the district level officers, or the institutional credit network, he can merely act as a trouble shooter on behalf of the Special Schemes Organisation, which normally prefers to deal directly with the Collectors and DRDAs, as also with the executive departments and banks.

For purposes of coordination with institutional credit agencies a state level Bankers Coordination Committee exists at the secretariat level and the Collector is chairman of the District Level Coordination Committee (DLCC) consisting of bankers and government officials. Currently Collectors are also administrators of the district Central Cooperative Banks and of the Zila Parishads, while Additional Collectors—Development are administrators of the primary land development banks. However, commercial bankers always look towards their own head offices or towards the Reserve Bank of India for directions and coordination is not always achieved. This is also because the orientations and objectives of bankers and government staff differ considerably. At Panchayat Samiti levels meetings of "Block Project Authorities" have been started by the government to increase communication between the local panchayat samiti DRDA and bank staff.

It is also pertinent to point out that the DRDA has, in addition to the Collectors, an Additional Collector Development cum Project Director (from either the IAS, RAS, or a person selected under special selection rules), a district planning officer, a credit planning officer, a project economist, a subject matter specialist (Planning), an accounts officers and an assistant project officer animal husbandry with a skeleton staff of clerks and statistical assistants. The development section in the Collectorate which is responsible for most of the Collector's coordination of developmental activities normally has one or at most two clerks with the Additional Collector Development as officer-in-charge and a deputy district development officer (who basically deals with Panchayat Raj work) as link officer in case of the absence of officer-in-charge. Famine work is normally dealt with by a separate section in the Collectorate having its own officer-in-charge and link officer (who may or may not be the same as for the development section).

IV

After a description of the structures, it is time to turn to problems. Before dealing with problems particularly relating to the structure of district desert development administration, it would be appropriate to deal with general problems of the rural development delivery system.

One is the fact that nowhere is significant decision making delegated to levels lower than the state level. Even relatively small purchases have to be made at head of department levels as per General Finance and Account Rules, leading to delays. Second is the tendency today to routinise campaigns so that unless a campaign is conducted nothing gets done in a routine manner, and a campaign for one programme automatically leads to the neglect of other work. The third problem is that field staff lack full and proper understanding of the programmes. Thus easy to perform tasks often get priority over vital and primary tasks. Another problem relates to the lack of zeal and motivation on the part of field functionaries who often have other gainful work to do and act as part-time government servants, or alternatively they act as a part of the local village elite[11] and thus only look after the interests of the dominant group in the area and not of the disadvantaged and the poor.

Detailed micro level planning has not yet become a part of India's administrative culture. Hence time spent on surveys and detailed planning is often considered a waste of time and not enough care is taken on fulfilling these preliminary tasks. Hence programmes well conceived on the macro scale fail at the micro level during implementation.

Often the local people (the beneficiaries) resist due to lack of understanding of the benefits being delivered. Efforts at increasing local peoples' participation may not always prove successful as instead of the deserving and the poor, the rich may capture the participatory mechanism and twist it to misappropriate the benefits[12]. The rich may also block development if they fear that it may upset the local power balance.

The practice of reviewing progress in terms of money spent and physical targets of benefit delivery achieved, rather than in terms of the long-term impact of the work also leads to the non-achievement of programme goals.

Based on these and similar reasons, a list of leakages points or stages in the delivery process where leakage occurs, can be identified as follow:[13]

1. Recommending and fixing norms and identification of both beneficiaries and benefits at policy making level.
2. Communicating guidelines of norms and identification of beneficiaries and benefits to operational staff.
3. Identification of beneficiaries in the field.
4. Wrong selection of benefits from the point of view of the beneficiaries or the community's needs.
5. Securing/selecting/constructing/approving the benefits (for example, selection and purchase of milk/animal or pump set, or faulty civil work construction).
6. Untimely (or unseasonal) and delayed providing of benefit to the beneficiary.
7. Lack of delivery of (i) knowledge necessary on part of beneficiary, and (ii) infrastructure facilities and other supportive pre-conditions for the beneficiary to be able to optimally use the benefit.

8. Wilful and forced transfer of benefits to another after receipt by beneficiary.
9. Premature reporting of delivery progress resulting in the delivery transaction remaining indefinitely incomplete as after making the report the delivery agency loses interest in the case.
10. Misappropriation by some outsider or repayment of loan by beneficiary causing his dept liability to increase and thus negating the value of the benefit.
11. Natural calamity causing loss of benefit.[14]

V

What about the problems inherent in the structure of district desert development administration.

No one can properly plan the work. The Special Schemes Organisation discusses new items separately with each executive department without knowing what other works that department itself is going to take up in the district. It can also not coordinate work of one department with that of other departments. Thereafter, even though the DDP work may be more deserving of priority, the departmental district level officer is likely to give priority to his own department's work rather than to the DDP work. Further, the lumpsum allocation of funds for sectors and departments made by the state planning department in conjunction with the Government of India is unalterable once made for a year and thus does not allow for flexibility in planning. Thus, what is likely to result is fragmented work by various departments. In Sikar district, the items covered under DDP by separate district level officers are: soil surveys, soil conservation, planning and execution and sprinkler demonstrations, afforestation, groundwater survey and test drilling, minor surface irrigation works, rural electrification, animal health and *gaushala* development, sheep breeding, and pasture development. It may be added that these works are equally suited to non-desert districts also. The Collector and DRDA also cannot coordinate the plan properly because the district level officers are not aware in advance what works, either departmental or out of DDP funds, are going to be assigned in a particular year. There are even instances when a corporation accepted money for a scheme proposed by it and later declared the scheme non-viable and offered to return the money.

Even meetings of the District Planning Committee are often a struggle between elected MPs and MLAs desiring to secure extra benefits of Minimum Needs Programme for their constituency and district level officers trying to stick to plans prepared at higher levels. Even at higher levels the decisions taken are not necessarily technical or administrative as the minister of the department may have intervened. Recommendations of district level committees are often ignored at decision making levels.

The Minimum Needs Programme envisages provision of select facilities to target villages at the least cost. Thus a road may be built to the village on the basis of least mileage to be constructed rather than on the basis of natural linkages of the village as discovered through a Rural Growth Centre Survey, or say the convenience for running a milk collection route by the dairy. This is an example of possible fragmentation in plan execution.

Another consequence of the structures and process is that the progress monitoring becomes exceedingly difficult. The funds transferred by the Special Schemes Organisation directly to executive departments are often spent centrally so the district level officers may have no idea of the progress. This may also often be the case of funds transferred by DRDA to district level officers, and by them to their superiors or co-equals. In such cases, progress reporting to the DRDA and by them to SSO is delayed.

Often resources and equipments meant for DRDA work in one district are temporarily diverted by the executive department to a DRDA or departmental work in the same or another district. In other instances, resources and equipments lie unused as departmental staff first undertakes departmental work and then turns to DDP work. Often getting administrative and technical sanction from both SSO and the other concerned department takes time. If staff vacancies exist they may be disproportionately higher in posts for DDP work than in posts for departmental work.

Collectors also have a tendency to pay more attention to tasks which the DRDA has to perform directly or through the revenue and development agencies, i.e., IRD, SFDA, TRYSEM, Antyodaya, and to leave the rest to routine reminders or to the quarterly meetings of the DRDA governing body. Only a few Collectors actually keep a detailed check on the working of the district level officers.

Technical supervision of the DDP works depends upon when circle or divisional officers visit the area on their normal tours.

The DRDAs are not aware of the state or sectoral positions and long-term perspectives and therefore cannot take the initiative in proposing locally suitable schemes to the SSO.

Coordinating of DDP and famine works may occur in particular cases, but just as departmental and DDP works are not always planned in a coordinated manner, similar is the case of famine vis-à-vis DDP works. The famine department is however now coordinating famine works and departmental plan work in advance. The DDP and DPAP strategy of preparing a shelf of project for implementation of famine works during famine years has failed due to two reasons. The long gestation period of the projects, and the pressure from peoples' representatives and the people to start other famine works besides those which form the shelf of projects. Pressure also builds up to undertake labour intensive works having low benefit cost ratios as such schemes having low B/C ratios cannot be taken up under other plan programmes (even though the B/C ratio norm for DPAP/DDP work is lower then for normal plan work).

The above may give a very pessimistic picture. And indeed things are not so bad as made out above. A lot of coordination does occur. But the point being made is that the scope for institutional improvements is vast; and that a lot of the coordination today is informally and personally achieved by particular Collectors, or Additional Collectors or officers of the SSO or of other departments.

VI

So what formal reforms are possible?

One is that some funds be set aside for each district for micro level planning in an integrated manner. Gujarat perhaps is experimenting with this[15] in that, while 80 per cent funds are to be used for planning as before, 20 per cent are being earmarked for integrated block and district level planning and implementation. Even for the 80 per cent, while the funds need not come to the DRDAs, this could be transferred directed to concerned executive departmental staff. The work of each department in each area should be so planned by the SSO that the works are mutually reinforce-ing and integrative in nature. For this more inter-departmental planning and review meetings shall have to be conducted by the SSO rather than the present trend of single department review and planning meetings. That such a strategy may not succeed is indicated by the experience of Rural Growth Centres Schemes in Rajasthan. The idea was that natural growth centres be selected and their hinterland identified after surveys. Thereafter for the centre plus hinterland detailed physical, financial and spatial planning was to be done. Works were to be executed as per the project report by various departments out of their plan funds with the SSO providing some additionalities through DRDAs. However, such integrated projects flounder at the time of execution because many works do not fit departmental norms for selection of works. Even if they do fit such norms, the concerned departments prefer to retain their autonomy to select work priorities and manner of doing work rather than to agree to undertake the work as required in the project report. Even if they agree to do the project work out of their own funds the time or spatial sequence of implementation may not be as envisaged in the project report. Great effort at the DRDA and SSO levels is required to be put in to coordinate implementation of a model RGC project which itself was watered down before the approval of the state government.

Thus, perhaps a more relevant solution would be to follow the CAD Chambal Rajasthan model whereby one or two or three districts are placed under an area development commissioner and where officers from various departments are placed fully subordinate to the commissioner and budgets for various sectors are also placed at the disposal of the commissioner, who can then coordinate both the planning and implementation of work in all sectors for the area without struggle with countless departments. For this to be successful not only the DDP or DRDA funds, but all developmental funds and all developmental tasks in the area would have to be made the responsibility of the commissioner who would prepare detailed project reports[16] and get them executed. Such plans would cover both the areas and the peoples' needs. Thus, functionally segregated departments would have to give way to areally integrated administrative units of government. The secretariat would also have to be reorganised accordingly. This should go a long way towards solving the special problems of the desert or similar special areas.

If simultaneously the local populace, rich and poor alike, can all be conscientised and organised to become partners in progress with the government functioning through the above proposed areal project units, then impact of the government programmes could perhaps be even greater.

NOTES

1. See, for example UN-ESCAP Bangkok, *Public Service Delivery System for the Rural Poor* (1980) and its earlier companion volume *Rural Development Administration in India: Some Emerging Policy Issues* (1979). The recent National Institute of Rural Development, Hyderabad book *Rural Development in India: Some Facets* (1979) indirectly discusses this, while *Dynamics of Rural Development* (1980) (mimeographed proceedings of a seminar at Mussorie, 23-25 May 1980) also underscores the importance that this issue has achieved in the minds of academicians and administrators alike.

2. And these descriptions of the role of the Collector mostly predate the emergence of special programmes like SFDA, DPAP, CAD etc. and the DDAs to implement these special programmes.

3. For the earlier role and later decline of Panchayati Raj, see (i) Government of India, *Report of the Committee on Panchayati Raj Institutions* (Delhi, 1978), and (ii) P.C. Mathur, "Performance of Panchayat Raj Institutions in Rajasthan 1959-74: A Critical Survey" in *Social Change*, Vol VII Nos. 3 & 4, September-December 1977.

4. This section has been developed from schemes earlier set forth in Rakesh Hooja, "Delivering Benefits to the Rural Poor: A view from Rajasthan", paper presented at the Government of India/ESCAP Working Party on improving public service delivery systems for the rural poor held at Delhi from 8 to 12 November 1979. Also see (i) Bhupendra Hooja, "Agriculture Administration in Rajasthan", in Satish Batra, *Administration in an Indian State* (Associated: Delhi, 1980), and Bhupendra Hooja and Rakesh Hooja, "Government Development Schemes and the Modernisation of Agriculture in India", forthcoming in *The Administrator*.

5. For details about the new agriculture extension agency, see (i) Bhupendra Hooja, "Agriculture Extension: A New Programme of Training at all Levels" in *Kurukshetra*, Vol. 27 No. 7, 1 January 1979, and (ii) Rakesh Hooja, "T & V in Rajasthan: An Attempt at Better Downward Communication through Agriculture Extension" in *The Administrator*, Vol. 23, No. 3 & 4, July-December 1978.

6. For the functioning of some of the these project agencies see (i) *Formulation and Implementation of Rural Development Projects* (mimeo) by Rakesh Hooja (Training Programme for Overseas Administrators, Government of India, Department of Personnel and Administrative Reforms, Training Division, Delhi, 1980); and (ii) Rakesh Hooja and S.R.

Mittal, "CAD and SFDA Loaning Problems and Procedures at the Field Level", *Financing Agriculture*, Vol. X No. 4, January-March 1979.

7. For the author's views on the institutional credit system, see his "Micro Level Credit Planning and the Rural Delivery-cum-Recovery System" in *Administrative Change*, Vol. VI Nos. 1 & 2, July 1978-June 1979.

8. See Government of India *Manual on Integrated Rural Development*, 1980.

9. See (i) Bhupendra Hooja, "Towards Antyodaya (Unto the Last)" *Rural Development Digest*, Vol 2 No. 3, July 1979, and (ii) Rakesh Hooja, "Antyodaya and the Possible ,Role of Non-Panchayat Officials in Executing Development Schemes", *Political Science Review*, Vol. 17 Nos. 1 & 2,, January-June, 1978.

10. For a description of how state planning is currently undertaken in Rajasthan vis-à-vis theoretical approaches, see (i) Kanta Ahuja (ed.), *District Planning in Rajasthan* (HCM RIPA, 1977); (ii) Rakesh Hooja, (a) "State Level Planning: The Rajasthan Case", *Indian Journal of Public Administration*, Vol. XXII No. 3, July-September 1976, (b) "The District as a Planning Unit: Style and Locus", *IJPA* Vol. XIX, July-September 1973, (c) "District and Regional Planning: Some Aspects with Regard to India", *Journal of LBS National Academy of Administration*, Vol. XX No.1, Spring 1975, (d) "Planning for Rural Development by Field Staff: A View from Rajasthan", *IJPA*, Vol. XXVI No.4, October-December 1980; and (3) Rakesh Hooja and Harsh Sethi, "Planning for Industrial Development in Rajasthan", *IJPA*, Vol. XX No. 3, July-September 1974. Also see R.P. Misra and K.V. Sundaram, *Multi Level Planning and Integrated Rural Development in India* (Heritage, Delhi: 1980), and Y. Venugopal Reddy, *Multi Level Planning in India* (Vikas: Delhi, 1979)

11. See *Dynamics of Rural Development*, op. cit. and Rakesh Hooja, "Economic Opportunities, Agrarian Structure and Rural Development in India", *New Quest*, May-June 1981.

12. Bhupendra Hooja, "Functional Approach to Panchayat Development—A Post Asok Mehta Perspective", *Administrative Change*, Vol. VI Nos. 1 & 2, July 1978 to June 1979, and Ramesh K. Arora (ed.), *People's Participation in Development Process*, HCM SIPA, 1979, have a more optimistic opinion of people participation.

13. This list has also been developed from an earlier paper of the author.

14. Some useful articles on rural development and the delivery system are: (1) D. Aurora, "Adaptation of Administration for Rural Development with Special Reference to Poor", *Behavioural Science and Rural Development*, Vol. 2 No. 1, January 1979; (2) Mohit Bhattacharya, "Administrative and Organisational Issues in Rural Development", Proceedings of the *22nd Annual Conference of Members of the IIPA* (IIPA: Delhi, 1979); (3) P.C. Mathur, "Administrative Realignments for Rural Development: Some Policy Issues in the Context of India", *Indian Journal of Political Science*, Vol. 40 No.1, 1979; (4) V.R. Gaekwad, "Management of Rural Development Programmes: Organisational Deficiencies and Strategy for

Improvements" in *IJPA*, Vol. XXXI No. 4, October-December 1975; and (5) R.P. Misra and K.V. Sunderam, *Rural Area Development" Perspectives and Approaches* (Sterling: Delhi, 1979)

15. As mentioned by K.V. Sunderam in his presentation on Planning Cycle Build From Below to Meet Local Needs at the Seminar on Concept and Approaches to Rural Development, Jaipur, 24-27 February 1981.

16. Such micro level planning could follow guidelines given in (1) *Manual on Integrated Rural Development*, op. cit.; (2) Prodipto Roy and B.R. Patil, *Manual for Block Level Planning*, Macmillan, 1977; (3) Government of India, *Report of the Working Group on Block Level Planning* (1978); (4) *Guidelines for the Formulation of Comprehensive Block Development Plans* (mimeo), Special Schemes Organisation, Jaipur, January 1980, and in mimeographed volumes of the UNIDS National Correspondence Seminar on Micro Level Planning jointly organized by Planning Commission, Government of India, United Nations Social Development Division, and Institute of Development Studies, University of Mysore and especially its volume *Guidelines for Grass Root Level Workers: Comprehensive Analysis and Planning at the Micro Level.*

6

A Methodological Note on Area Development Programming: Some Reflections on Western Rajasthan, 1975

HARSH SETHI

There is a well developed and growing body of literature on the methods of underdevelopment planning in two particular areas: the techniques of economy-wide planning and the methods of project evaluation. The analytical basis for both the categories of planning methods lies in the theory of resource allocation. There are well defined concepts like feasibility, efficiency, and optimality in the context of resource use in an integrated economy, and in the theory of resource allocation, economists do have some useful analytical results—of course under certain assumptions—regarding technology and behaviour.

The techniques of regional planning, however, need to be different from those of economy-wide planning, primarily because the set of constraints demarcating a regional economy is usually different from those demarcating a national economy e.g., movement of goods is less restricted, no trade policy instruments are available etc. Also the scale of operations at the regional level is to small—and the nature of inter-dependence among actuaries in geographical space is such as to rule out project evaluation. Projects can no longer be considered in isolation.

Still another important point of departure arises in the case of a development problem of a backward area. A typical backward area is characterised by slacks in the use of resources—because of lack of complimentary inputs or lack of marketing facilities. Thus the effectiveness of a regional development programme should be viewed not only in terms of scarce resources but also in terms of its induced effects on locally available idle resources.

Many difficulties are faced by regional planners—regions vary from each other in terms of economic and geographic characteristics—and each distinction requires its specific technique. Therefore, in formulating techniques of regional planning one must be clear about the kind of region one is planning for and how the economy of that region is linked with the rest of the economy. The region shall be assumed to be relatively small in comparison to the entire economy so that the development plan of the region does not alter the picture of relative scarcities in the entire economy.

Assume that the region is a backward agricultural region. We must start with the development potentials—agriculture, animal husbandry, and allied activities. First, explore the production possibilities of the region with reference to these activities. The production possibility set depends upon the availability of primary factors and other inputs for production—which are the labour force and the land area. Of course in dynamic programme they cannot be taken as given because of possibilities of migration and land reclamation. What is the agro-climatic condition, what about availability of other inputs like water, fertilisers, agricultural implements etc.? Let X^A be a vector of outputs of this set of activities which a development plan can make possible to produce within the region.

Now consider the non-agricultural set of activities, e.g. transport and social infrastructure. But the demand for these depends not only upon the agricultural sector but also on trading and import-substituting activities of the region. The identification of tradeable commodities depends upon resource endowments, transport advantages, technical characteristics etc.

The best procedure to search for candidate industries is by looking into the current pattern of imports into the region for possible import substitution and by examining the needs for processing the primary produce of the region. Alongside you also have to look into allied service facilities like warehousing, marketing etc. Let these activities be called X^n.

The set of non-tradeable basic activities like transport, power, health, education etc. may be called set X^b.

For preliminary consistency, obtained by applying input-output norms—the final goods coming out of the total production system of the region is:

X^A Agriculture

$X = X^a$ Industry = gross production rector

X^b Services

Let A = Input-output matrix of the commodities in the X-sector. Then (I−A) X=F=final goods vector.

Elements of F for non-tradeable commodities have to be non-negative, and match the estimates of regional demand.

Now, we have to decide which regional development of the type X is the best to initiate. Here one cannot apply the standard inter-industry techniques, as industrialities and scale problems creep in.

Thus the method to use is to prepare a few alternative programmes of the type X, evaluate each of these complexes, and choose the best one. The question is how to choose X.

The Model

The ideal method of evaluating such a regional development programme is to measure the net effect of the programme on the inter-temporal development objective of the whole economy. Compute the discounted present value of net benefits—in which every individual item of costs and benefits is measured as the social willingness to pay for it in terms of the weighted sum of the aggregated consumption of the economy and other distinct social objectives. This has first to be done in a static framework. All costs and benefits are expressed it terms of annual flow variables. Outputs and current inputs correspond to periods when capacity utilisation is maximum as in the case of "Social Evaluation of Single Investment

Projects", all inputs and outputs should be measured in terms of corresponding shadow prices.

We have:

$(I-A)\ X = F_L + F_N$

where F_L=Local demand for the output of these sectors, and

F_N=Net export of these sectors.

The Revenue—p FL+p.FN.

p=Local demand price vector.

p=Inter regional trade price.

Let k=vector of capital—output ratio

r=Rate of return

d=Annual rate of depreciation

Then $(r-d)KX$ = Annual capital charges for the investment programme.

Annual wage cost = WLX

where W=Shadow wage rate vector

L=Vector of labour output ratios.

Further, the production of X shall require current material inputs from outside the programme set.

Let C be the vector of exogenous cost/unit of output

: Net annual social profits=

(p. FL+P.FN)−(r−d)KN-WL.X-C.X)

Note: We have not discussed any distributional criteria.

(2) Social profits are equated to welfare.

Now one has to measure the induced effects. We have to identify only those activities which are likely to be induced by this programme. To measure the net indirect benefits one should deduct from the value of gross output, costs due to scare resources, used by these induced activities.

If for the E_K category of such induced activity—we call YK the gross income and ZK the opportunity cost of resources used, then aggregate indirect benefits is:

$E_K(YK-ZK)$.

Thus the aggregate benefit of the development programme is

$$p.FL+P.FN-(r-d)K.X.+WL.X.+CX)+E_K(YK-XK)$$

Choose the programme with the highest value.

The main purpose of presenting the above format is to underline the inter-disciplinary nature of efforts required to formulate a development programme. Specialist knowledge about the production possibilities of land need to be combined with those regarding the various urban-industrial sectors to constitute of viable rural-urban mix of the regional development programme. The advantage of presenting a combined framework of project evaluation and input-output analysis is to invite feasibility and consistency tests from the various sectoral and regional specialists

II

When one talks of the desert areas of western Rajasthan and plans for them, the attempt is not to present an elaborate model of a complex on industries and infra-structural facilities coming up in the desert. Even a cursory look at the area will reveal this to be a vain hope. Both in the absence of water, existing metropolitan centres and the lack of financial resources in the state, the important thing is not to radically alter the present structure but rather to maintain the cultural balance of the area better. Thus, the entire sense of suggestions should be reviewed keeping in mind that the object of the presentation is to increase the income of the poorer sections and to maintain them there and stopping the influx of *refugees* into towns every time a severe drought hits the area.

A reading of the Techno-Economic Survey of the NCAER on Rajasthan reveals simultaneously a few interesting possibilities and many severe constraints to development.

The most notable thing is the absence of water. The Rajasthan Canal does not service this area. Rainfall is a miracle and about underground water resources we know very little as yet. Unless a major programme of desalination of sea-water and pumping it to this area in huge quantities is carried out the visions of major centres of civilization are indeed only visionary.

The soil suffers from much erosion—either sand or rocky stretches—lie all around you. The heat and the lack of vegetation would probably kill anyone other than the native and of course sheep, goats and camels.

Potential for agricultural development is limited. Desert soil and a lack of water permit only dry farming. The only crops grown are millets, grams and pulses and some primary cereals—wheat, barley and maize. This area is a permanently deficit one and probably will continue to remain so.

The hope lies in animal husbandry—sheep, goats and camels. Rajasthan accounts for about 20 per cent of the sheep, 16 per cent of goats and about 60 per cent of the camel population of the country. Probably the livestock population exceeds the human population.

Very little has been done to utilise this resource in the state. Most of the people in the area of concern, tend livestock. Our objective should be to make this more profitable for them. There is a Sheep and Wool Department in Rajasthan; there is also a factory in Jaipur producing combs and cutters as part of the wool shearing machinery. Both need to be given more encouragement and to increase the scale of their operations. At present the factory faces a situation of deficient demand because private parties don't do this, not because they see no profitability in the venture, but because they don't know where to train their operators. It is therefore incumbent upon the government to train private operators also, alongside with government employees.

A vigorous co-operative sector, in the hinterland of Rajasthan with modern machinery and trained staff, could do wonders with the sheep. Much more wool can be realized from existing livestock.

Another thing that needs to be done immediately is the starting of large-scale sheep farms—preferably in the state sector. This will help push up the quality of the sheep population of the state—and also permit scientific experimentation.

Besides the woollen industry—much scope exists for the development of the mutton industry. Here the goats also come in. Small-scale meat packing industries in the towns of western Rajasthan have potential of a brisk business. The state farms can be the chief suppliers to such factories.

To really develop this industry the fodder situation in the area needs to be improved. This again demands state involvement, otherwise we shall continue witnessing the mass migration of livestock every year into neighbouring states.

We should not forget our camels. Besides being excellent draught and transport animals, much can be done with their skin, meat, and bones. Local enterprise given help and a little push will take care of all this.

The government very often talks of mineral development in Rajasthan. Unfortunately, the government considers only large deposits and thus only large-scale programmes. The possibility of small-scale mining done in the co-operative sector, working with mainly surface one, does not seem to appear profitable to them. My talks in the Geological Survey of India Office, revealed the technical possibility of indulging in such small-scale mining.

The major problem seems to be organisational. But as long as our objective is not so much to extract large quantities of ore but mainly to provide stability to the local population—at a no profit no loss basis—or if necessary a small subsidy—I am of the opinion that the scheme is worth trying. The social gain, taking in account the externalities involved, is very much greater than the costs concerned.

The third stage of link in our development programme is the promotion of rural handicrafts. Anyone who has seen the products in places like the Pushkar Fair can have no doubt about the quality and the potential market of these goods. The government does have a fairly extensive handicraft organisation but this seems to be working mainly with artisans living in the urban centres. Again the problem is primarily organisational. Rural Cooperative Societies should be able to penetrate into the hinterland and permit the truly beautiful handicrafts of rural Rajasthan to be appreciated all over.

My entire presentation seems to avoid any over-reorganisation of development schemes. The framework of planning suggested is very loose and involves maximum amount of indigenous participation. To make a stagnant area come alive—without heavy loss of capital inflow and all forthcoming contradiction—requires an institutional framework conducive to such a scheme.

A major problem in this area concerns transportation. Excepting a few roads connecting major cities and some border roads made by and for the armed forces, the rest of the area is not serviced at all. Population pressure does not seem to warrant building of 'pucca' roads—which would enrich only the contractors—unless of course a road building programme is carried on as part of the rural works or famine relief programme. Camel transport could solve the problem. Transport is needed not only for communication but also for the movement of water supplies, mineral ore, finished products etc. New designs of camel carts with rubber wheels have been produced in the state—and with a large camel population already present—we would do well to use it.

Let me briefly re-state some of my reflections upon the problems:

(1) An intensive survey of the NCAER type should be conducted for the districts of Bikaner, Barmer, Jalore and Jaisalmer.

(2) An attempt should be made to describe how the present economy hangs together and special emphasis should be placed on examining the potential of increase in: (a) dry farming, (b) animal husbandry and allied industry, (c) rural handicrafts, (d) small scale mining.

(3) Industries on which to concentrate:

(a) Sheep and wool (b) Leather (c) Meat

(d) Rural handicrafts (e) Light minerals

(4) Decide on the type of infrastructural facilities i.e. transport system.

(5) What should be the new organisational set-up.

The government's recent interest in this area and the extensive work done on:

(1) water resources;

(2) social conditions;

(3) new mineral ores, is a very favourable sign.

I think an intensive study team comprising of different disciplines, spending some time in the desert and semi-desert area, carrying out surveys under the guidance of the "Central Arid Zone Research Institute", would be a welcome addition to the series of explorations already going on.

7

Combating Desertification in Drylands: A Planned Approach

R.P. SINGH

Of the 628 million people inhabiting the world's drylands, 72 per cent live within semi-arid zones, 27 per cent inhabit in the arid zone, and one per cent live in the extremely arid zone. The fate of approximately 14 per cent of the world's population is thus linked with the productivity of drylands (Anonymous, 1977).

Dryland ecosystems are fragile all over the world. Overgrazing and unwise cultivation practices have rendered these ecosystems prone to desertification. Today, about 50 and 78 million people are directly threatened by desertification.

The consequences of desertification are well understood. Desertification leads to loss of productivity of the resource-base, deterioration of the health and well-being of the people at risk, and desertification-induced migration and resettlement.

Controlled land use of dryland ecosystems offers the most certain return. The key of such control lies in maintenance of an adequate plant cover. An array of strategies and techniques have been developed over the years by the people inhabiting the drylands, to cope up with high degree of rainfall variability.

Inter-annual rainfall variability, so characteristic of the drylands, calls for contigent planning, with an in-built flexibility and diversity, to combat desertification. Of all the production systems followed on drylands, the one related to crop production is perhaps the most vulnerable to weather aberrations. In principle, land systems vulnerable to erosion should not be cultivated. Withdrawal of land from agricultural is the most prudent policy in case where farming involves too great a risk. Land systems of extreme western Rajasthan fall in this category.

Management of Drylands Ecosystem

On a long-term basis, soil and moisture conservation measures hold the key to salvation of tropical drylands. The measures have a long period of gestation and are labour and capital intensive. They are best pursued by the community rather than by individuals. The synergistic effects are obtained when these measures are planned and executed on a watershed basis. Such measures include:

(i) establishment of wind breaks and shelter belts,

(ii) strip cropping of grasses and crops,

(iii) stubble milch farming,

(iv) water harvesting and its recycling,

(v) installation of small water and sediment-retaining dams, check dams in gullies, and

(vi) planting of quick growing cover on devegetated areas and gullies.

Bare dryland fallows serve as loci to desertification-deflation soil in dry years and sheet/gully erosion in wet years, ultimately leading to reduced biological productivity of the dryland eco-systems. Elimination of the bare fallows and maintenance of cover crops/stubbles, coupled with contour farming procedures, is the surer way of imparting resilience to the system. A favourable micro-climate aids in the process of resilience. The soil's capacity to retain nutrients, organic substances, and a high infiltration capacity are the prerequisites. An integrated nutrient supply system holds the key to providing the basic elements of resilience.

On a short-term basis, however, judicious crop planning, matching the soil moisture availability and weather aberrations, is helpful in imparting stability to crop production on dryland eco-systems.

Crop Production Strategy Based on Weather Aberrations

A study of weather aberrations on a long-term basis is essential for formulating a sound crop production strategy for drylands. The strategy for abnormal years has to be highly flexible and location-specific. Dryland Research Cente at the Central Arid Zone Research Institute, Jodhpur, Rajasthan (India) is engaged in planning and formulating research and development strategies for normal and abnormal rainfall years for the last ten years. An analysis of rainfall distribution patterns obtainable at Jodhpur, Rajasthan for a 80-year period (1901 to 1980) has revealed the following aberrations:

(a) Early onset of monsoon
(b) Late onset of monsoon
(c) Early recession of monsoon
(d) 'Breaks' in monsoon at the seedling stage, and at the flowering/grain formation stage of dryland crops.

Strategy for Early Onset of Monsoon

The south-west monsoon sets in around 1 July at Jodhpur. In 27 years out of 80 years, the monsoon had set in early to mid-June (Table 1). In situations like these, the Dryland farmers are caught unawares. Experience at the Central Arid Zone Research Institute, Jodhpur, has shown that cereal crops like pearl millet give good account of themselves when sown early in the middle of June. Sesamum sown early does as well. Grain legume crops, in particular *mung* beans, castor beans, and *moth* beans, etc., when sown early, exhibit a tendency to put up excessive vegetative growth, thus resulting in low harvest index. So is the case with early sown castor beans. Early sowings are sometimes fraught with risks of crop failures, should a long spell of break/moisture stress follow sowings. Despite this probability, the risk is worth taking.

Table 1

Analysis of Rainfall Pattern (1901 to 1980)

Situation	Years
Early onset of monsoon	27
Late onset of monsoon	20
"Breaks" in monsoon	
(a) at seeding stage	18
(b) at flowering and grain filling stage	24
Early recession of monsoon	26

A fodder-cum-grain production system could also be planned for such a situation. In the event of the pre-monsoon shower soaking the soil profile deep enough (40 to 50 cm), fodder crops like pearl millet or sorghum could be sown with advantage. The crops can be harvested for fodder for 45 to 50 days after sowing. Pearl millet has a better capacity to tolerate atmospheric drought than sorghum. A short-duration legume crop (*mung* beans or moth beans), raised for grains, could follow the cereal fodder crop. The system has a capacity to give 120 to 150 q/ha of green fodder, besides 4 to 5 q/ha of grain legumes, within a period of four to five months Alternatively, castor beans could be sown for fodder early in the season, followed by late sown castor beans for seeds. This production system has a potential of giving about 150 q/ha of green fodder and 6 to 8 q/ha of castor bean seeds.

These production systems have three basic advantages:

(a) the farm labour is gainfully employed in pre-monsoon months, which otherwise remain underemployed or unemployed,

(b) the soil loss occurring as a result of wind erosion is minimised/ eliminated because of early crop cover in the wind erosion-prone month, i.e., June, and

(c) inclusion of a legume in the system ensures a cheap supply of nitrogen and maintenance of soil fertility.

Rainfall quantum and distribution permitting, even three crops (pearl millet fodder, *mung* beans, mustard) could be raised on drylands, thus providing adequate crop cover almost all the year round.

Two practices common to traditional desert agriculture that need attention are pre-sowing operation called 'sood' and minimum tillage.

Both these practices display lot of wisdom with regard to conservation of resources. In the first case, selective cutting and clearing of shrubs and bushes (the operation locally called 'sood') is carried out in June as a pre-sowing measure. Bushes like *zizyplus* are not uprooted, but are carefully maintained on the field. These bushes provide the much sought-after 'pala' (dry leaves of the bush) in the lean months of fodder availability. It is very nutritious and palatable and hence precious. More importantly, these bushes provide anchorage to the otherwise loose soil prone to wind blowing. In the second case, i.e., minimum tillage, the soil is disturbed as little as possible. Ploughing, therefore, is a taboo, unless necessitated by hard pan beneath, restricting water intake. Provided the two components of traditional desert agriculture, enunciated above, are kept intact, the crop production systems suggested for early onset of monsoon, will pay rich dividends, both in terms of productivity and maintenance of soil fertility.

Strategy for Late Onset of Monsoon

There have been 20 years out of 80 (Table 1) when the onset of monsoon was delayed by two to three weeks. A strategy comprising the following components is suggested for such situations:

- Alternate crops and varieties
- Dry sowing
- Pre-sowing seed treatment
- Transplanting (in case of pearl millet)
- Early and complete weed control

Table 2

Comparative Performance of Dryland Crops Sown Late in the Season

Grain yield (q/ha)

Years	Pearl millet (HB3/BJ-104)		% Change over NS	Sesamum (T-3)		% change over NS	Mung bean (S-8)		%change over NS
	NS	LS		NS	LS		NS	LS	
1972	F	1.5	—	F	1.9	—	F	2.4	—
1976	20.4	F	−100.0	3.4	F	−100.0	12.7	3.7	−70.0
1978	17.9	7.1	−60.3	2.9	3.0	+ 3.4	10.6	1.6	−84.9
1980	5.2	2.4	−53.8	0.8	0.2	−75.0	6.3	1.9	−69.8
Years	Moth bean (Jadia)		% Change over NS	Castor bean (FS277)		% change over NS	Sunflowers (EC58414)		%change over NS
	NS	LS		NS	LS		NS	LS	
1972	F	3.0	—	F	2.3	—	F	3.1	—
1976	7.1	2.6	−63.4	10.9	5.0	−54.1	3.3	0.9	−72.7
1978	0.8	1.4	+ 75.0	9.8	6.8	−30.6	2.4	3.1	+ 29.2
1980	2.8	1.8	−35.7	5.8	3.5		—	1.1	—

Note: 1. 'NS' stands for normal sown crops (end of June or first fortnight of July), while 'LS' stands for late sown crops (end of July or first fortnight of August).

2. 'F' stands for failed

Experience gained over the years at the Central Arid Zone Research Institute, Jodhpur, suggests that cereal crops (pearl millet, etc.), when sown late in the season, result in low and unremunerative grain yields. Yield reductions over normal sown crops are drastic. Grain legumes (*mung* bean, *moth* bean, castor bean) and oilseed crops (castor bean, sunflowers, etc.), on the other hand, give reasonably good yields (Table 2). Between castor bean and *moth* bean, the former gives higher yields over the latter, under late sown conditions (Table 3). Not all varieties of the promising grain legume and oilseed crops, listed above, perform well under late sown conditions. For example, varieties S-9 of *mung* beans, K-11 of cowpeas, FS-277 and 2470/12 of castor bean, give better performance under late sown conditions, than the varieties S-8 of *mung* beans, FS-68 of cowpeas of normal onset of monsoon.

Table 3

Comparative Performance of Castor Beans and Moth Beans Sown under Normal and Late Sown Conditions

		Normal sown	Late Sown
Castor bean (D.S.)	1979	4.8	2.5
	1980	—	—
Moth bean (T 18.)	1979	4.5	0.2
	1980	2.8	1.5

"Dry sowing" is not much in vogue in desert districts of Rajasthan. Two reasons may be advanced for this. First, preponderance of pearl millet in the cropping system, and second, sandy loam and loamy sand soils occurring in the region. Both the soil types are prone to soil surface crusting. Pearl millet being a small-seeded crop, does not have the capacity to emerge out of the crusted soil. Sesamum is not dry-sown either, precisely on this account. Preliminary observations recorded at the Central Arid Zone Research Institute, Jodhpur, have shown that *mung* bean, *moth* bean, castor bean and peanuts, could be sown with advantage in a dry soil, in anticipation of monsoon. These seeds in castor bean and peanuts in particular, have an amazing capacity to 'break' the crust and emerge out as clean and clear as the normal sown crops. Dry sowing gives an early start to crops, which take advantage of the moisture available early in the season. Higher seed rates than normal, and deeper seedings (within the limit of maximum seeding depth), often cover up part of the risk involved in the system. Early weed control is a pre-requisite to the success of dry-sown crops.

Soaking of seeds having hard seed coats, like sunflower, castor etc., in water for 24 hours prior to sowing, proved effective in giving higher coefficient of velocity of germination. Sowing of soaked seeds thus gives an early start to dryland crops. The practice will be found all the more useful under receding soil moisture conditions.

Should it become necessary to take pearl millet late in the season for considerations of meeting the family's requirements of staple foodgrains, transplanting is the only way out. Transplanting, though labour intensive ensures a good crop of pearl millet, despite not-too-congenial an environment. Transplanting 21 to 25 days old seedlings of pearl millet, raised in a nursery by a watersource, late in the season (fourth week of July to first week of August), gave consistently higher grain yields over the late sown direct-seeded crop of pearl millet (Table 4).

Table 4

Comparative Performance of Direct Seeded and Transplanted Rainfed Pearl Millet

Treatment	Years			Mean
	1977	1978	1979	
Direct seeded	22.9	5.3	2.2	10.6
Transplanted	27.7	15.3	3.3	15.4
Percent increase over direct seeded pearl millet	21.0	88.7	50.0	52.5

Early weed control, within 30 days in the case of *moth* beans and 40 days in case of castor beans, resulted in higher seed yields that weedings done later in the life cycle (Table 5). Yield reductions are very if weeds are allowed to compete with crops till their maturity (Table 5) Early and complete weed control is all the more important in case of crops sown late in the season, for weeds compete with crops sown late in the season, for weeds compete with crops for moisture, a commodity which gets scarcer with the advancement of the season.

Table 5

Yield (q/ha) of Moth Bean and Gowar Bean as Influenced by Different Timings of Weeding

	Moth bean			% reduction in yield	Gower bean (2470/12)			% reduction in yield
	79	80	Mean		79	80	Mean	
Weed infestation for first 10 days	3.9	1.1	2.5	—	3.5	2.7	3.1	—
20 days	3.7	0.8	2.4	4.0	3.1	2.3	2.7	13.0
30 days	2.8	0.3	1.7	32.0	2.9	2.0	2.5	19.4
40 days	2.5	0.5	1.5	40.0	1.6	1.9	1.8	41.9
50 days	1.8	0.4	1.1	56.0	1.7	0.2	1.0	67.7
Weed infestation till maturity	1.3	0.2	0.8	68.0	1.0	0.2	0.6	80.6

Low plant populations set out in wider row widths, in case of crops like *mung* beans sown late in the season, may help achieve higher yields and moisture use efficiency (Table 6).

Table 6

Seed Yield of Late Sown (13 August) Mung Bean as Influenced by Varying Plant Populations

Plant Population (lakh/ha)	Seed yield (q/ha)	Yield attributes		Moisture use efficiency (Kg ha/mm)
		Pods plant	Seed wt plant (g)	
5.7	3.4	9.0	1.5	2.6
2.9	3.6	9.7	2.3	2.8
1.9	3.9	10.9	3.0	3.2

Strategy for Early Recession of Monsoon

Dryland crops are left in the lurch when the monsoon recedes earlier than normal. The normal date of recession of monsoon at Jodhpur is 8 September. In 26 years out of 80, the monsoon receded early (third week of August or so).

Very little could be done with regard to modification of plant architecture through stripping off of unproductive (contributing to photosynthesis) lives, at this stage. Moreover, the procedure is cumbersome and impractical over large acreages. However, the following strategy is suggested to cope up with such situations:

(i) adjustment of plant populations according to soil moisture availability,

(ii) removing the most vulnerable component out of the inter-cropping systems,

(iii) recycling of rainwater, harvested and stored in a pond earlier in the season,

(iv) rationing,

(v) use of organic mulches, and

(vi) use of anti-transpirants.

Thinning out of alternate plants in the row in case of mild moisture stress, and removal of alternate rows in the case of severe moisture stress, are attempts to adjust plant populations according to soil moisture availability. Removing the most vulnerable component in an inter-cropping system is yet another version of adjustment or mid-season correction. Evidences are available to show that with limited moisture, in a population of 80,000 plants/ha, all the sorghum plants were earless. But when the population was brought down to 30,000 plants/ha, earless plants were considerably less and at 20,000 plants/ha the emergence and grain setting were satisfactory (Anonymous,1976). Similarly, encouraging results were obtained when the sorghum component in the sorghum-sunflower inter-cropping system was removed when the drought set in.

Providing a life-saving (supplemental) irrigation to dryland crops, undergoing moisture stress is the surest way of stabilising yields of dryland crops. Rain water harvested and stored in a pond earlier in the season could be recycled for providing supplemental irrigation. One supplemental irrigation (5 to 7 cm), given at an appropriate growth stage, resulted in grain yields increases of the order of 13 to 37 per cent over the unirrigated control. In a drought year the effects of supplemental irrigation were even more spectacular with 337 per cent yields increase (Table 7).

Exploratory studies carried out at the Central Arid Zone Research Institute, Jodhpur, have indicated that early sown pearl-millet crop could be rationed 45 to 50 days after sowing, giving a green fodder and grain yield of 220 to 250 q/ha and 8 q/ha respectively.

Table 7

Effect of One Supplemental Irrigation (6-7 cm) on the Yield of Kharif Crops Grain/Green Pod Yield

Crops	1977		% increase	1978		% increase	1979		% increase	1980		% increase
	I_0	I_0		I_0	I_1		I_0	I_1		I_0	I_1	
Pearl millet												
Direct seeded	22.9	24.4	6.6	5.3	6.0	13.2	2.2	9.6	337.0	—	—	—
Transplanted	27.7	29.7	7.2	15.3	21.0	37.3	3.3	5.7	72.8	3.3	5.7	71.9
Castor beans	4.7	5.2	10.2	—	—	—	—	—	—	—	—	—
Cluster beans (vegetable type)	—	—	—	—	—		—	—	—	48.4	64.5	33.4

Note : I_0 stands for no irrigation, while I_1 stands for one supplemental irrigation.

Evaporation losses could be cut down through the use of organic mulches. In case of sandy loam and loamy sand soils, the surface mulches are not of much avail, for the upper soil layer, once getting dry, serves as a mulch itself. On a saturated or a nearly-saturated soil profile however, application of 6 to 8 tons/ha of *bajra* husk mulch resulted in higher productivity and moisture use efficiency, which was equivalent to the one achieved by the use of polyethylene mulch (Table 8). Use of anti-transpirants could sometimes be helpful in mitigating part of transpiration losses, thus resulting in higher moisture use efficiency.

Table 8

Grain Yield and MUE of Pearl Millet as Influenced by Different Mulches

Treatment	Grain yield (q/ha)		MUE (kg/ha/day)	
	1972	1973	1972	1973
Control (no mulch)	9.3	31.4	8.1	9.1
Dust mulch	9.6	26.3	8.4	7.9
Wheat straw	10.4	25.0	8.8	7.4
Bajra husk	11.8	29.9	10.8	9.1
Polyethylene	11.8	29.0	10.2	8.6

Strategy for 'Breaks' in Monsoon

'Breaks' in monsoon are a common phenomena experienced in dry climates. These may occur either at the seedling stage or at flowering/grain filling stage in the life cycle of dryland crops, the latter being more common than the former (Table 1).

Breaks in the early stages of crop may lead to mortality of seedlings. The only way to recoup the loss is to resort to gap-filling, preferably through transplanting, wherever possible. Early seedling vigour, induced through judicious fertiliser use at the time of seeding may help overcome the much dreaded calamity—the mortality of seedlings. Interestingly, shoot-fly incidence on sorghum was found to be much more devastating on unfertilised plots than on fertilised plots (Anonymous, 1976).

Breaks late in the season, besides being more common, are more devastating in as much as either very poor crop yields are obtained or at times even crop failures results. In order to combat this eventuality, the rain water stored in the pond should be recycled for supplemental irrigation. At least it will serve as a crop-life-saving device, if not more. A staple food crop thus saved is like a family (inhabiting the drylands) saved from starvation, migration and resettlement. It is foolish to talk of economics of water harvesting and recycling, particularly when it comes for saving a family or a community from starvation. Farm ponds should be planned and provided on a water-shed basis in order to be more effective and viable. There is no surer way than water harvesting and recycling for saving dryland ecosystems from desertification and consequent devastation. Land use, according to capability, of course, is yet another way.

Strategy for Normal and Good Rainfall Years

Principles of contingency planning are as much application to good rainfall years as to bad years. Unlike bad years, plentiful moisture supply should be capitalised in good rainfall years. Good harvests obtained in normal and good rainfall years can pave the way to stability. They can help overcome the poor harvests of bad years.

The following strategy should click in years and situations of plentiful moisture supply:

1. Diversifying the resource-base of the dryland ecosystems by introducing more stable components like forest trees, horticultural crops, pasture grasses, etc. into the system.
2. Establishing 'community-nurseries' of fruit and silviculture plants to cater to the needs of the community.
3. Establishment of 'fodder banks'.
4. Diverting more and more water to 'khadins', so as to bring more area under cropping.
5. Harvesting inevitable run-off and storing the same in a pond for future reuse.
6. Utilising surplus water for leaching out salts from salt-affected soils.
7. Exploring the possibilities of a double crop system and a more productive inter-crop system.
8. Judicious use of inputs like fertilisers and pesticides.

Strategy for Systems other than Crop Production

Pastoralism is a common vocation of dryland arid ecosystems. Resilience of the pastoral ecosystem is better ensured through diversification of herd species, and grass/legume species. The former strategy is based on the simple anology and premise that "single species are often less effective at keeping the range in good condition and less productive than the variety of grazers on the same land".

Reducing numbers of the stock, like thinning of crop plants, is a sound strategy for dryland pastoral ecosystems. Culling of diseased or unproductive animals is yet another. Maximum productivity comes from healthy herds.

Besides limiting the size of herds, herdsmen spread their stock thinly over large areas, resort to rotational or deferred grazing, and provide additional watering points. In case of nomadic pastoralism, mobility is the chief weapon to fight against a harsh environment.

A simple strategy for increasing the productivity of wild pastures lies in:

(a) removal of coarse, dead stems, thus allowing the sprouting of new succulent shoots, and

(b) allowing a quick low-temperature fire, thus helping in bringing up new green shoots and removing old touch ones, thereby helping in speeding up nutrient cycling.

Location and management of water points is the key to the sustained productivity of the pastoral ecosystems. Water points, like farm ponds, should be planned and provided on watershed basis and according to the strength of the stock.

It must be realised that dryland ecosystems are highly fragile. Mismanaged or carelessly managed dyrland ecosystems lead to desertification. Besides, following the proven soil and moisture conservation procedures, contingency plants should be followed according to varying weather aberration situations. Emphasis should be laid on low cost monetary inputs or technology, as farmers inhabiting these drylands are poor.

8

Grassland Management for Desert Reclamation & Rangeland Development

L.D. AHUJA

Biophysical conditions of desert areas have been described by different scientists and development workers elsewhere. The land utilisation pattern in arid regions of Rajasthan (Table 1) reveals that area under cultivation is higher (61.3%) in Zone 'B' receiving average annual precipitation of 300 mm and above; while it declines (31.4%) sharply in Zone 'A' with annual rainfall below 300 mm. In districts like Jaisalmer (with annual rainfall below 200 mm), the area under cultivation during normal years of rainfall is about 5 per cent; hence most of the uncropped land is mainly grassland, a source of livestock production for the local economy. The shortage of forage for existing livestock population amounts to 42.0, 35.0 and 35.1 per cent in Zone 'A', 'B' and total arid zone respectively. The situation is alarming during years of scarcity.

The uncropped lands and those falling under land capability (class V) and above are sources of devastation of nearby fertile agricultural lands through erosion hazards. Rangeland or grassland may be defined as a landform where agriculture cannot be sustained on sound footing due to hostile agro-climatic conditions. These are best suited for use as rangelands, a good source of livestock sustenance. Due to continuous misuse, these lands are deteriorated resulting in economic losses. Studies were therefore conducted under the CAZRI, Jodhpur from 1959 odwards on 52 rangeland sub-station under different agro-climatic conditions. These sub-stations have been reduced to 12 since 1969. The findings are summarised as under:

Table 1

Livestock Population in Different Agro-climatic Zones in Arid Districts of Rajasthan and its Pressure on Land

Agro-climatic zone	Total area 000' per ha.	Area (000'ha)		Total livestock ACU (millions)	Land available in ha/ACU			Per cent shortage of forage during normal years
		Under cultivation	Under grazing		Total	Cultivable	Grazing	
Zone 'A'								
Mean annual rainfall below 300 mm	11,498	3,610 (31.4)	5,785 (50.4)	2.043	5.62	1.76	2.83	42.0
Zone 'B'								
Mean annual rainfall above 300 mm	9,361	5,741 (61.3)	2,777 (29.7)	4.690	1.996	1.22	0.59	35.0
TOTAL	20,859	9,351 (44.8)	8,562 (41.1)	6.733	3.09	1.38	1.27	35.1

Figures in parentheses indicate percentage.

ACU = Adult Cattle Units

Rangeland Development

As per an ecological survey of the country, there are five distinct grassland covers in India and that under arid region which falls under *Dichanthium-Cenchrus-Lasiurus* cover. Within this, several grassland types exist. Under the *Dichanthium-Cenchrus-Lasiurus* main cover of western Rajasthan, there are also 12 distinct grassland types in Luni basin (11,000 sq. km.).

Different condition classes of rangelands in arid regions and their carrying capacity are defined as under:

Condition class of grassland	Forage production kg/ha during normal years	Carrying capacity adult cattle units (300 kg/ body) wt. per 100 ha.
Excellent	1,500 kg and above	25-30
Good	1,000 and above	20
Fair	750 kg	17
Poor	500 kg	13
Very poor	200 kg or even less	0-6

Forage yield could be substantially increased if existing "poor" rangelands are improved and brought under "good" and "excellent" condition class. In addition to increased production, regular source of social menace and nuisance, viz., erosion hazards, drifting sands which devastate nearby villages, fertile agricultural lands, infrastructures etc., could be minimised. Techniques of improving rangelands are discussed below.

Selection of Sites

It will be desirable to take up grassland development work in the beginning in areas with easy accessibility and water facility so that local people can visit the places frequently and realise the value of such developmental works. Cooperative attitude of local people at such sites is essential.

Fencing

In this part of the country, no development project involving land can be successful without proper fencing (protection from biotic factor) due to heavy livestock pressure. Among the various types of fencing tried, the one with barbed wire with angle iron posts is by far the most effective and economical in the long run, with proper maintenance lasting over 25 years.

The present cost of angle iron post and barbed wire fencing is about Rs.10 per running metre. Smaller the size of the block, greater is the proportionate cost per hectare. The cost of such fencing per hectare in blocks of 40,100,200,500 and 1000 hectare works out to Rs.660, Rs. 450, Rs.300, Rs.200 and Rs.130 respectively. Maintenance cost of such fencing is very low.

Forage Production

Efficient grassland management aims at maximum utilisation of forage without undue interference with growth and vigour of important fodder and other useful species. With protection and grazing the native rangeland on the carrying capacity aiming at 70 percent forage utilisation level, the forage yield increased and was practically double in 2-3 years time. The data on forage yield under different condition classes of rangelands for six years are presented in the following table.

Table 2

Forage Yield (Air Dried kg/ha) in Different Type of Rangelands in Arid Zone, after Protection and Controlled Grazing

Type of rangeland	Year of Observations						
	1959	1960	1961	1962	1963	1964	1965
Poor	337	430	837	688	642	770	885
	(9)	(21)	(18)	(12)	(8)	(9)	(9)
Fair	552	873	1,065	869	817	1,074	835
	(10)	(22)	(21)	(17)	(11)	(10)	(10)
Good	—	704	1,068	1,523	1,263	1,480	1,163
		(4)	(6)	(8)	(9)	(8)	(8)

Detailed studies under different agro-climatic condition reveal that the forage yields increased as rainfall increased from western to eastern Rajasthan. It is not the absolute precipitation during a year which affects forage production from rangelands, but the pattern of distribution of rainfall during the year influences the productivity from different plant species. Forage production was also found to be dependent on the habitat, particularly the soils for obvious reasons. Production was low on shallow gravelly habitat, and highest on heavy deep soils.

Grubbing of Unwanted Bushes

Unwanted thorny plants like *Lycium barbarum, Balanites aegyptiaca, Acacia leucophloea, Mimosa hamata* etc., which hinder the growth of grass species and are troublesome to grazing animals should be grubbed out mechanically followed by application of herbicide 2-4-5 T (trichlorophenoxy-acetic acid) immediately after cutting away serial parts of bushes. Debushing of such obnoxious species in dense intensity followed by reseeding with *Cenchrus* sp. resulted in dominant cover of useful species like *Cenchrus* sp. and high forage yields of 1.1 t/ha during the first year, which increased to 4 t/ha during the next year at Pali.

Top Feed-cum-Shade Trees on Grasslands

Trees provide top feed, shade to grazing animals and timber and fuel on the range. Top feed in the form of leaves and pods (rich in proteins and minerals) provide valuable nutrition to the livestock during lean periods of the year, thereby increasing the secondary productivity.

Suitable top feed species could be planted along the boundaries, inspection path and approach roads of the compartments of the rangelands in single or double rows. They will also act as wind breaks. Species like *Acacia nilotica* var. indica and *Prosopis juliflora* grow well on mounds of corewall fencing to provide additional protection also to the property. Trees should be planted in staggered manner in rangelands. A 'good' rangeland may have about 30 to 50 such trees per hectare.

Some tree species permit growth of grass species under them. Contribution of ground storey (grass component) under *Prosopis cineraria* is highest (over 1t/ha/yr) as compared to that under *Acacia Senegal.*

Soil Conservation

On range lands with shallow soils and rolling topography different soil conservation measures, viz., contour furrows, contour bunds and contour trenches have been tried. Contour furrows with 929 sq. cm. cross section 61 cm width, 22.6 cm depth, spaced 8 to 10 m apart increased forage yield by 638 per cent (average of 10 years) and proved superior in all respects to the other techniques. Cost of such works is about Rs. 150 per hectare and could be recovered through increased forage production in 3-4 years period.

In case of land at foothills, the flowing *nalas* should be harnessed with *pucca* masonary and earthen bunds across the slope to serve as water spreading devices. Gulley and nala plugging should be carried out wherever necessary.

In extremely arid region like Lawan (annual rainfall below 200 mm, Jaisalmer district) rangeland with gravelly shallow soils, contour furrows (in between trenches and core wall 400-500 m apart), trapped sands on which *Lasiurus sindicus* established and forage yield increased from 659 kg/ha (mostly herbs and low perennials) to over 2 t/ha (mostly high perennial (climax) species *Lasiurus Sindicus*.

Stabilisation of sand dunes: In arid zones, shifting send dunes are active, except during monsoon. They threaten the existence of nearby village, towns and fertile lands. If stabilised and managed properly, these mobile sand masses offer a potential productive site for afforestation and grassland development. For stabilisation and rehabilitation of sand dunes, four distinct processes are involved, viz., (a) protection, (b) mulching, (c) afforestation, and (d) planting of seedling or root slips of drought hardy perennial tussocky grass species. The cost of stabilising the dunes presently works out to Rs. 600 to 800 per hectare.

Rangelands with salinity: First step in reclaiming such land is to drain out excessive salts. Digging trenches cum mounds with dimensions of trench as 4'+2'/2×3' or 1.22+0.61/2 m × 0.914 m at a distance of about 300 meters (depending upon the slope) helps to drain the excessive salts. In such systems, grasses like *Spprobolus helvolus* (kharada) a highly, nutritious and palatable grass species comes up naturally. As the reclamation of salinity progresses the production increases, and in fairly well leached patches, *Dichanthium annulatum* comes up luxuriantly. Such managed grasslands could yield 1.0 to 1.5 tonnes of air dried forage per hectare per year. It is advisable to sow or plant salt resistant tree species like *Acacia nilotica, Prosopis juliflora, Tamarix articulata, Salvadora persica, Acacia tortilis* etc. on the mound of trenches to stabilise them and save them from being damaged during rains.

Reseeding Grasslands

Natural succession of superior perennial species of grasses in arid and semi-arid regions is a time consuming process. Hence reseeding with suitable perennial grass species having high nutritive value and yielding high tonnage under rainfed condition, suiting different agro-climatic tracts is the only solution. Reseeding of grasslands involves the following:

Selection of suitable species for reseeding in different tracts: Dichanthium annulatum gives high yield in heavy soils with annual rainfall above 380 mm. *Cenchrus ciliaris* and *Cenchrus setigerus* produce good tonnage on well drained soils under medium to low rainfall, while *Lasiurus sindicus* gives encouraging forage yield on sandy soils with low precipitation (300 mm and below). *Panicum antidotale* performs well on medium textured soils with annual rainfall of 250 and mm above under protected conditions, while *Sehima nervosum* yield good forage on hilly terrain.

Reseeding: It involves proper soils working, sowing of appropriate grass species and after care of resultant cover.

In regions having annual precipitation below 300 mm, pelletting of seeds of *Lasiurus sindicus* is recommended. Sowing just before expected monsoon season is advantageous. Next best is sowing immediately after first effective monsoon shower. It is advisable to resort to sowing in lines about 75 cm apart.

Weeding: Natural rangelands are a potential source of weeds for obvious reason. Juvenile growth of high perennial grass species is slow. Reseeded plants of perennial grass species meet with stiff competition from high population of seeds. Therefore, such reseeded rangelands need at least two weedings during their first year of establishment.

Production: As per the data concerning increase in forage production due to reseeding under different condition classes of rangelands, increase in forage yield due to reseeding was from 30 to 122 per cent in 'poor' and 29 to 107 per cent in 'fair' rangelands.

Optimum forage yield: Optimum forage yield in well established plots of different high perennial species under different rangelands was studied and results discussed by Ahuja and Mann (1974, 1975).

Upto 3.6, 4.7 and 8.9 tonnes of forage per hectare were obtained in established stand of *Lasiurus sindicus, Cenchrus* sp. and *Dichanthium annualatum*.

Hilly and gravelly terrain: In hilly tracts, at foodhills and gravelly areas reseeding is generally not a practical proposition. Such areas should be brought under the purview of Forest Act and State Soil Conservation Act, and protected and vegetation allowed to come up.

Saline soils: In highly saline soils, reseeding with high perennial grass species has not given encouraging results, and on such land form systems lower perennial grass species like

Sporobolus helvolus performs well. It is quite nutritive and possesses high production potentials.

Rangelands with climax vegetation: Extensive areas in low rainfall regions of Jaisalmer, Bikaner and Barmer districts, covered with high yielding nutritious grasses like *Lasiurus sindicus*, remain underutilised due to shortage of potable water in the region or are subjected to overgrazing where watering facilities exist. In such areas grazing could be regulated (i) during monsoon season when water requirements of domestic animals are lower and some run off water is available in open ponds for animals and stock owners, (ii) by dividing area into suitable blocks for practicing rotational grazing, and (iii) surplus forage may be harvested and stacked for future use.

Fertilisation

The nutrient content and production potentials of forage species on 'good' rangelands is quite high (crude proteins about 7 to 12 per cent). Therefore, for optimised production, it is essential to provide adequate nutrients in the soils. Placement of fertilisers like sulphate of ammonia and urea at 20 to 30 kg N/ha have given encouraging results.

Fertilisation of 'good' rangelands in zones under rainfall below 300 mm and above 300 mm with 20 to 30 kg N/ha increased forage production by 1,155 kg/ha (31.9%) and 1,369 kg/ha (34.1%) respectively.

Utilisation

The grasses (forage) on rangelands could be used by harvesting, and/or grazing. As harvesting of forage coincides with other agricultural operation and is also a costly affair, it is desirable to utilise the herbage on the rangeland through judicious grazing. Animals while grazing add nutrients to the soil through their droppings and also break the encrustation of fine silt on the surface of soil which hinders the infiltration of rainwater and thus aid towards betterment.

Morphological, physiological, growth status and chemical constituent characters of the available plant species on the range determines their utilisation by different species of farm animals.

Studies on grazing in the rangelands under different intensities of grazing stress and deferred rotational grazing indicate that (i) even with heavy intensity of grazing the rangelands, production from perennial species like *Lasiurus sindicus, Cenchrus* sp. is not affected for short periods of 3-5 years, (ii) forage production is not affected under rotational versus continuous grazing for 3-5 years, and (iii) however range productivity is affected by rainfall and its distribution pattern.

Livestock Production

Growth of yearling heifers grazing on carrying capacity of rangeland is highest during monsoonic period (July to October) when forage on the range is nutritious and green and it is of the order of 12.7 to 17.5 kg/animal/month (animals belonging to the local stock owners). It is favourably comparable with livestock growth production on good plane of nutrition. Thereafter growth rate slows down and from December onwards animals just maintain their body weight. Growth on yearlong basis was 88 to 96 kg/animal/year. The growth of animals gets doubled on yearlong basis if the cattles are given supplemental feeding to meet DCP and TDN needs from December to June. Cost of such concentrate feeding is Rs. 1.40 per kg increase in body weight based on 1966-68 prices index.

Performance of sheep and goats on the rangelands has been satisfactory. Studies on effect of mixed grazing with sheep and cattle; and sheep and goats on range performance are under progress.

Optimum growth of yearling ramlambs (sheep, during August-December) was 107.3 gms/day/animal on saline soils/rangeland; 195.5 gm/day/animal on rangeland under *Cenchrus-Aristida* cover, and 145.3 gms/day/animal on rangeland with *Eleusine* cover.

Optimum growth on yearling basis of yearling ramlambs of Jaisalmer breed, Marwari breed and Chokla breed of sheep was 27.8, 27.4 and 19.3 kg per animal respectively.

In general growth of animals (livestock production) is influenced by rainfall and its distribution and soil type which finally affect the vegetation complex.

Grazing behaviour and preference palatability of different farm animals for various range plants during different seasons has been studied in details.

Formulae to evaluate body weight of cattle through measurement of heart girth for estimation of feed needs have been evolved.

Water requirements of different species of farm animals on the range have been worked out and 'tanka' device to collect rain water for stock and human consumption on the range has been evolved (Ahuja, 1980).

Development

Between 1961 and 1979, 78 tonnes seed of perennial grass species viz., *Cenchrus* species, *Lasiurus sindicus* and *Dichanthium anmulatum* have been collected and distributed to different development organisations of various states for reseeding to improve rangeland production.

Effect of adapting rangelands improvement techniques is substantial (Table 3). In addition to remarkable increase in carrying capacity, improvement in vegetation and land form, there is minimisation in social evils and nuisance of erosion hazards.

Table 3

Carrying Capacity of Rangelands under Different Conditions Adult Cattle Units (ACU) per 100 ha.

Particulars	Average annual rainfall	
	300 mm and below	Above 300 mm
Present carrying capacity (ACU)	4-8	12-20
Carrying capacity after protection (ACU)	8-20	24-30
Carrying capacity after soil conservation reseeding etc. (ACU)	20-30	40-48
Increased carrying capacity due to improvement practices (3-1)	16-22	26-28
Increased yield of forage tonnes (2.5 tonnes/ACU)	40-55	65-70
Value of increased forage at Rs. 200 per tonne	8,000-11,000	13,000-14,000

Some *constraints* of these are discussed below:

(a) With limited rainfall, specially in arid region, sowing period of agricultural crops and reseeding of rangelands is very small (10-15 days in a year) and these coincide with each other.

Human beings are more interested in agricultural crops with the result that rangeland development work has not been able to make substantial headway.

(b) Rangeland development work is mostly concentrated on land under class V and above which have rough terrain, are subject to erosion hazards and therefore have disturbed vegetative cover. Their cost of development is higher than that of well-developed agricultural lands.

(c) Major inputs for development are fencing and reseeding. Due to ceiling on land, cost of fencing is high. It is inversely proportional to size (area) of the land. Farmers neither have financial resources for it nor facilities to procure it. Reseeding also involves substantial expenditure. Farmers need financial assistance for it. Paucity of grass seeds is one of the main hurdles.

Conclusions

To stop social and economic evils resulting from erosion hazards from the present desertic areas, following suggestions for development of rangelands may be considered.

(i) It may be taken up as a national problem and tackled through government organisations/cooperative societies.

(ii) Chairs for range management may be created in universities/agricultural universities for teaching under graduate and post-graduate classes, to train them for transfer of technology.

(iii) Fodder banks may be instituted in each panchayat to meet with the problems and needs of years of scarcity.

(iv) For proper utilisation of rangelands, adequate soil conservation law may be enforced shortly.

(v) Sufficient grazing fees for grazing animals on improved rangeland may be levied to enable the recovery of cost of development and maintenance.

9

Public Health Problems in Desert Environment

B.S. CHAUHAN

In folklore deserts are described as forbidden waste lands. So, the people living in these areas were also forgotten and written off till few years back. Now with our exploitation of natural resources and threat of population explosion, man has been forced to turn towards deserts in the name of development for their stores of solar energy, minerals, wind power, live stock products and land banks.

Desert development is proposed to attack on desert problems through the strategy of improving the economy of the area through developing a sound infrastructure and focusing on farm development activities, with the objective of optimum utilisation of land, waters human and livestock resources in the area.

Having viewed the list of activities in both DPAP and DDP, it has been observed with a mixed feeling of sorrow and surprise that 'health' component' is missing, though optimum utilisation of human resources is clearly mentioned.

In the past also, even with good intentions, health was forgotten, as health is taken for granted. Even in charters framed after World Wars I and II health was included at a very belated hour. Development plans are mostly prepared by those who are affected by 'personal encounter-physician' system rather than community health system' in their day to day life; naturally that influence operates while the plan is drafted.

In these programmes, the main emphasis has been put on economic development and health has been left to wait till economic development is completed. Without careful examination of who would be benefitted from it, it is assumed that economic growth will spread throughout and will take care of health as well. Economic development is a long drawn process and health is an urgent necessity, which cannot be allowed to wait longer. Health and development are two faces of the same coin, they are interdependent and synergetic in action; health is a resource to development.

The concept of human welfare and well-being in its broad based sense is clearly defined if linked to the concept of 'quality of life and total development'; this has to be brought about by exchange of opinions and experiences, with policy makers, national leaders, and those technicians and experts who are close to the aspirations of their people and are implementing technical programmes in health and other sectors of development.

To achieve self-reliance a country needs to develop its own resources—human and material—at all stages and phases of the development.

Since health and socio-economic development are inter-dependent and these two are inescapable issues, these must be exploited simultaneously. They have a reciprocal feedback effect that varies in its results according to the different historical, political and social conditions involved.

Development planners have, both in the past and now, in desert development plan attached little importance to health services. If they considered them at all, they tended to judge them by whether they led to per capita increase in production. However, most health services in developing countries do not bring about any lasting improvement in health status, and even in developed countries differences in health standards cannot be explained by the amount invested in services such as doctors and hospitals.

By the 1940s, Sweden had achieved much lower mortality rate than nearly all its European neighbours with few number of doctors per thousand population. In the Netherlands maternal mortality rates were low even when vast majority of birth took place at home; on the other hand the USA spent more on its health services per head than any other country in the world, but had poor mortality rates.

Most of the effort and expenditure on health in developing countries has been on curative services, mainly to the urban population. In many countries only 10 to 20 percent rural population have reasonable access to health services. So health services in conventional pattern in developing countries have not improved the health of people. The most skilled curative services may make only a temporary impact on health, if basic causes of ill health are not remedied at their origin in rural areas.

The planners should not judge the contribution of health services to national development by short term effects on out put but by long term effects on quality of life. For instance, if survival rate of infants in the developing countries is raised, which can be done without much doing with poverty, future adult population could be produced with much lower use of resources. Much healthier labour force, could be produced with same resources, at the same time leaving women with time and energy to make larger contribution to the economic and social well-being of their families.

Socio-economic planning with health should be seen as complementary and not competitive. The role of health administrator in the past had been that of an advocate for planning and spending on health services. He has rarely made a substantial contribution to the nation's total health development plan. The health administrators has a vital role to play in explaining the possible effects of various development plans on human health. The diseases can be prevented and overcome by careful design of the plan.

The health administrator will be alert to the proposed changes in the population of cattle, sheep, poultry and other animals, in view of the need-to protect man from diseases transmitted by animals.

To have a sound foundation, planning should move from bottom up and not top down, so the health must be incorporated at the infrastructure level. Local communities should thus be the first to be involved in the planning process. Diagnosis of what is wrong and when what is needed is known would save time and money. So health administrators can help to identify changes that are of special importance for health and forewarn the planners before they are caught unaware. He can be helpful in the survey of social conditions, monitor the effects of earlier plans from health point of view, identify the deterioration in conditions for various sections of the population and provide an early warning of the impending crises.

With respect to contrast between the substantial fall in death rate following the introduction of tuberculosis chemotherapy and the long steady previous decline of death rate, one can say with rough calculation that without drug technology, it would have taken forty years to reach this stage. The infant mortality rate can be reduced with the help of technology without much change in poverty; an outstanding example is tuberculosis, which has been recognised as one which has classically flourished in poverty. Yet drug iso-mazid worked all over the world, even in situations that permitted no real improvement in circumstances of living. When introduced in poverty areas the rate of fall in death rate has been identical with that elsewhere. Before 1947 steady fall in death rate of tuberculosis was due to raised standard of living rather than efforts of doctors. But after 1947, when drug use came in there was a steep fall in death rate.

So health plans should be inducted in the overall integrated development plan at the infrastructure level without waiting for economic development. This will make it, scientific, rational, economical and benificial to the planner and the people.

As is known, climate exerts a great influence on the health and socio-economic development of man. Public health problems in desert environment are peculiar and different than in non-deserts areas due to the peculiarities of deserts. For better study of desert health problems a concept of 'desert medicine' has to be accepted, after that they have to be considered under two heads:

(a) After development (after interference)

(b) Before development (without interference)

(a) After Development

In agriculture sector, irrigation activities will induce the following diseases:

1. Transmitted by mosquitoes:

   (i) Malaria

   (ii) Filaria

(iii) Anthropoid borne:

(a) Yellow fever

(b) Dengue fever

(c) Haemorhagic fever

(d) Equine sencepbilitie

2. Transmitted by Simulicum fly:
   Monchoecriceasis (riverblindness)
3. Transmitted by Testse fly:
   Sleeping sickness
4. Borne by snails:
   (i) Schistosomiasis (urinary interestinal)
   (ii) Pragonimiosis (lung flukes)
   (iii) Distomatosis (liver flukes)
5. Borne by fresh water:
   Guineaworm
6. Contamination
7. Diseases passed through rodents:
   (i) Rat lute fever
   (ii) Plague
8. Snakebite (snakes which come to eat rats)
9. Pesticides
10. Fertiliser
11. Vegetations produced
12. Insects surviving on these plants, etc.

Like this every sector of development will have its group of likely diseases, for which the health administrator has to be ready.

(b) Before Development

Diseases in the deserts are mostly due to geogens (geographical factors), rather than pathogens. Following are the geographical factors that may cause diseases:

1. High environmental temperature:

(i) Heat cramps

(ii) Heat expansion

(iii) Heat pyrexia

Children and old persons are more prone to heat stress. Infection and strenous exercise hasten the process.

2. Low environmental temperature:

At times during winters, under-nourished, unsheltered, ill clothed, lying in open, on sand dunes, exposed to blowing sometimes moist winds affects a person with generalised body chilling effect.

3. Sweating disorders—in deserts predispose to heat stress:

(i) Blocked tubules

(ii) Hypohydrosis

(iii) Transverse myelitis

(iv) Sympatheclomy

(v) Leprosy

(vi) Goat

(vii) Myxoedema, scurvy etc.

4. Solar Radiation:

(i) Sun burns

(ii) Erythema

(iii) Hugebullae

(iv) Lupus erythemetosis

(v) Rodent ulcer

(vi) Drugs inducing photo sensitivity.

Sulphonamides, sulfonylures, chlorthiazides, tetracylines.

5. Sand storms:

(i) Respiratory diseases

(ii) Silieosis

(iii) Tuberculosis

(iv) Trachoma

(v) Corneal injury

6. Water loss:

(i) Mild voluntary ch-dehydration

(ii) Low physical and mental efficiency

(iii) Opium addiction

(iv) Barkish water

(v) Flourosis

In deserts, diseases are presented with disguised and distorted clinical picture. Laboratory findings have to be unmarked.

Disguised form:

(i) Dehydration: presents with high fever, neligium and convulsions—mistaken as meningitis.

(ii) Sand particles when inhaled—are mistaken as post operative chest infection.

(iii) Vitamin 'A' deficiency cases of cattle and man: present as acute diarrhoea.

(iv) Heat pyroxia: as post operative infection.

(v) Sodium and waterloss—acute—mistaken as acute low back ache.

Distorted picture:

1. Bleeding shoulder—continues to have capillary oozing, already dehydrated—so B. P. goes still low.

2. Chest diseases may be aggravated.

3. Vitamin 'C' deficiency cases, bleed more after snake bites.

4. Bitots spots appear early in vitamin 'A' deficiency.

5. Flourosis: may appear with even permissible fiouride content—due to greater water metabolism in hot climate.

Such a health programme will provide a community diagnosis and treatment and economise health services in sparsely and scattered populated area, and make it health for the people, by the people of the people.

Droughts, famines and flash floods are frequent natural catastrophes in deserts. A comprehensive plan to deal them must be made out in advance with full knowledge of unnatural and unconventional foods eaten during famines. Other immediate problems like opium addiction, malnutrition, snake bites, flourises etc. should also be tackled.

Health planners for desert regions should have stayed and worked and preferably have been born in these areas. They will be able to work nearer to the aspirations of the people and will be able to involve the community in solving the problems.

II

Water and Watershed Development

10

Water in the Development of Desert Area and the Origins of Rajasthan Canal Project: A View from 1980

KANWAR SAIN

The Great Indian Desert encompasses half of western Rajasthan and extends over parts of Gujarat and Haryana. It covers nearly 2,18,000 sq. kms. Topographical survey carried out in 1954, showed that the desert was spreading at the rate of one mile (0.8 km) per year for the last 50 years and encroaching upon about 15,000 hectares of fertile land every year. The consequences of using land beyond its capacity are always adverse. This is specially true of the arid lands, as the balance between land, water and animals is more critical. When the balance is upset, the consequences are more severe because recovery from the effects of overuse usually takes longer in arid zones.

The National Commission on Agriculture in its report on desert development (1974) observed: "The main task of the desert rehabilitation and development has yet to be faced. In the meantime, the desert is deteriorating rapidly. Rise in human population led to an increase in the livestock population to support it. This created increasing pressure on the vegetable resources of the desert leading to their rapid depletion. The deterioration has reached a stage when unless it is effectively and expeditiously checked, there will be further deterioration at an accelerated pace. The longer the delay in its rehabilitation, more difficult, more expensive and more uncertain, will the development become".

Water is the most effective tool in the development of desert areas. This is proved by experience at many times in the world. Egypt would have been a desert, but for the Nile waters, brought from distant sources outside Egypt. Similarly, Punjab and Sind would have continued to remain deserts, but for the water of the Indus rivers, brought from outside their geographical areas.

Gang Canal

The Gang Canal provides an ample proof of the effectiveness of water in the development of Rajasthan desert. It is said that soon after the assumption of power by Maharaja Ganga Singh in the year 1899, the Bikaner state had a severe famine which took a heavy toll of life. The tragedy left a deep impression on the mind of the young ruler. To avert repetition of such a catastrophe, the Maharaja strove hard to get a share from the Sutlej Valley Project, when it was being drawn up in Punjab. The Bahawalpur state vehemently objected to supply irrigation water from Sutlej at Ferozepur to Bikaner, a non-riparion state. The framers of the project were also not sure if irrigation would be successful in the Bikaner desert. The Maharaja appears to have been so convinced of success that he did not deter from pledging the scantly resources of the state to make his financial contribution to the cost of the project. The Punjab government agreed to construct the headreach of the Gang Canal in its territory for about 100 miles and to operate it for all times at the cost of the Bikaner state.

The development of irrigation under the Gang Canal was remarkably rapid. Irrigation increased by about nine per cent annually. During the decade immediately after independence, development has been strikingly rapid with an increase of over 36 per cent. The project envisaged an intensity of irrigation of only 63 per cent; the actually intensity achieved in 1958-59 was as high as 92 per cent. An important public health benefit of irrigation was the provision of good drinking water. Prior to the canal project, people had to drink blackish and impure water.

Thar Project

The instance may be cited of the development of the Thar desert area, now in Pakistan. It consists of nearly five million acres.

Tradition, supported by geological evidence has it that the Indus formerly flowed down the middle and deposited huge quantities of sand and later shifted its course to the west. Sand dunes arise from fine material blown from coastal and desert region by wind. The vegetation consisted of low brush and scanty grass. There are no indications of earlier occupation as are found in certain parts of Rajasthan desert.

The question of developing the Thar area was first considered in 1870. No action was taken till a colonisation bill authorised the construction of a canal to irrigate the Shamlat area in 1901. Nothing was again done till a detailed project was prepared after field investigations by the author in 1936. It was difficult as the entire area to be commanded was full of sand dunes. I sent a few random samples for analysis and evaluation to the Irrigation Research Institute at Lahore. Dr. Mackenzie Taylor reported that the soil was unfit for irrigation as it had no nutrition and will not be able to grow any crop. This naturally caused a lot of concern.

The Governor of the Punjab called a special meeting of all the chief engineers of irrigation department to take a decision to stop further survey work on the Thar Project. I was specially called from the field. On my way to Lahore, when I was passing through an area which was proposed to be irrigated, I found that near some wells, an excellent crop was growing. I cut the matured jawar *bhuttas* from these sites and wrapped them carefully in a newspaper, giving the exact location of the fields and the villages in which the wells were located. I took these to the meeting at the Governor's House. Dr. Mackenzie Taylor, on the basis of the soil, expressed his opinion that large investment was not justified to bring such land under irrigation. Before the Governor recorded the decision of the meeting to stop further work on the investigation of the project, I was asked to make my comments, if any. I opened up the paper bundle and put the various samples of *bhuttas* on the table giving the location of the fields from which the *bhuttas* of jawar had been taken by me. Then I commented: "These jawar *bhuttas* are as good as grown in Lyallpur colony on the Lower Chenab Canal. As a practical irrigation engineer, it is my view that if the area is given moisture at the time when the crop needs it, it would be able to grow a good crop. In view of the scanty rainfall in the area, the Thar Project in my opinion is fully justified." This brought the meeting to an abrupt close and the work on the Thar Project remained continued.

The work on the Thar Project was begun on 1939. It is said that 2,50,000 of refugees from India in 1947 were settled in the Thar area. In August 1949, the Thar Development Authority was established to be responsible for the development of an area comprising 83,45,000 acres. Side by side with agricultural development, industries such aș sugar, cotton textile, wool, cement were also established.

The authority of Tribal Development Agency originally covered 1.5 million commanded by the canal for irrigation. In 1953, parts of the 3.5 million acres not commanded by the canal were proposed to be covered by well irrigation.

Rajasthan Canal

A preliminary report on the Rajasthan Canal was sent by the Government of Bikaner state to the Government of India in October 1948. According to the report, about 6.9 million acres of the Bikaner state and about half a million acres in the north-west corner of Jaisalmer state could be brought under irrigation by flow and lift, limiting lift to a maximum of 150 ft. By flow alone, about 2.8 million acres could be irrigated. The main sources of water indicated in the report were: (i) the flow of Beas river water, (ii) surplus Ravi water, (iii) about 5000 cusecs of Chenab water through the contemplated Mashu tunnel, (iv) storage dam on the Beas river and (v) dams on the tributions of the Jamuna.

When the project was considered in the Central Water and Power Commission, it was dubbed as a wishful dream. Only when I joined the commission in 1949, proper investigation of the project was taken up. The project was sanctioned in 1957-58 and the work started in 1958-59. In order to try out the crops which would grow best in the area on being irrigated, two experimental farms were set up—one at Barore and the other at Hanumangarh. At Barore, after the soil had been leached to reduce salinity, 60 acres were planted with such crops as green manure crops and various varieties of rice, cotton, jowar, bajra and ground-nuts. The cotton crop proved highly successful. Three varieties of rice were found suitable for the soil conditions. At Hanumangarh, 125 acres were marked off into eight test areas. Here the bajra crops proved excellent. Good results were also obtained with cotton and jowar.

Only a limited lift area was included in the 1957 project. The total area included in the project was 3.7 million acres. Thus, a large area of the desert which could be covered by lift still remains deprived of irrigation facilities.

Culturable Area of Rajasthan Still Requiring Irrigation

The culturable area of Rajasthan, according to agricultural statistics, is 587 lakh acres. Out of this 44 per cent is annual sown. The area which received irrigation till 1961 was only 11.3 per cent of the sown area and less than 5 per cent of the total culturable areas. The Rajasthan Canal will cover only 37 lakh acres. These figures will convince any one that the most important problem for the development of Rajasthan is shortage of water.

The water is important not only for purpose of irrigation, but it is even more important for supplying the domestic needs of the population in areas where rainfall is scanty.

Sources of Water

Waters of the three Indus rivers allocated to India by the Indus Treaty viz., Sutlej, Beas and Ravi, will be fully utilised by Punjab, Haryana and Rajasthan Canal. In future water could be had from storage on the Jamuna river and by diversion from the Ganga river system. This possibility is discussed further in detail. The second likely source is the groundwater from the waterlogged areas in the Punjab and Haryana, and the third source is weather modification.

Weather Modification

In India, from ancient times, attempts have been made to induce the clouds to give more rain. In recent years, rain-making experiments have been extensively carried out in Australia, USA, USSR, and France. Other countries which conducted these experiments include Canada, Japan, Israel, Pakistan and India. Results of these rain-making experiments have indicated that it is possible to increase the rainfall by about 10 to 20 per cent in situations where rain would otherwise have occurred. In this connection, it is worthy to note that the amount of precipitable water over Jodhpur in July is 5.5g. The corresponding values in Delhi, Nagpur, Bombay and Trivandrum are 6.1, 5.5, 6.2 and 6.6g respectively. The rainfall at Jodhpur is low because of adverse atmospheric conditions. The presence of the large volume of dust in the upper air affecting the radiation regime has been held responsible for the low rainfall in spite of good precipitable water in the air. It should be possible to alter this atmospheric condition by cloud seeding. Modern efforts on weather deficiencies of condensation nuclei in the clouds.

Experiments with 'warm cloud-seeding' in India indicated an increase of 20 per cent. If, by artificial means, it is possible to increase precipitation by 10 to 15 per cent, it may pay off the cost of the entire cloud-seeding operation. According to Bowen's calculations, the cost is less than one penny per acre per year. The financial return from this additional rain produced as a result of cloud-seeding, depends upon the purpose for which is applied. For instance, if we consider hydro-electric generation, if there is one per cent increase in the precipitation, the additional expected gain pays for the whole cost of cloud-seeding experiments. Therefore, if 10 per cent increase is achieved, then it is possible to get a handsome benefit. In purely agricultural operations, about five per cent increase in the rainfall is required approximately to pay off for the cloud-seeding operations. If, however, at crucial crop growth period rainfall can be increased the physical yield of the crop will be much more rewarding.

In view of the promise that experiments so far carried out in weather modification hold, it is suggested that Indian Meteorological Department should conduct systematic research in the Rajasthan desert to determine the most effective method for inducing more rain in the cloud formative conditions in this area by scientifically designed and vigorously checked experiments and evaluate the amount of precipitation resulting from such experiments.

Under the conditions existing in Rajasthan, I feel very optimistic that some additional water would be secured by weather modification techniques.

Additional Water from Punjab and Haryana

As irrigation is being extended, water logging is being experienced in certain areas in Punjab and Haryana. Also, by improved techniques in agriculture, at least 10 per cent of water could be saved. After a couple of decades, it should be possible for the Central government to persuade these states to spare this water to the desert areas in Rajasthan in the overall national interest.

Water from National Water Grid

The total surface water resources of India have been assessed at about 1440 million acre feet. Of this, only 250 million acre feet have been used so far. At the most 500 million acre feet can be beneficially used only by transferring water from surplus river basins to the areas having water shortage.

About six years ago, the Ministry of Irrigation envisaged interlinking the Ganga with the Kaveri involving very high lift. Recently, the Ministry of Irrigation and experts of the Central Water and Power Commission have formulated a national perspective for use of surplus water. This envisages building of as many storage dams as possible to conserve the water during monsoon months and transferring this to drought prone areas which experience difficult-ties of getting even drinking water. The broad approach adopted is to respect the existing uses to meet reasonable needs of basins.

The scheme comprises two major components. The first is the Himalayan Rivers Development, which envisages construction of storage reservoirs on the Ganga and the Brahmputra and their principal tributaries in India and Nepal alongwith inter-linking canal systems to transfer surplus flows of the eastern tributaries of the Ganga to the west, apart from linking of the Brahmaputra with the Ganga. Besides providing irrigation to an additional area of five million ha., the generation of about 30 million KW of hydropower, it will provide substantial flood control in the Ganga-Brahmputra basin. The scheme will benefit not only the states of the Ganga-Brahmputra basin, but also our neighbours such as Nepal and Bangladesh as well as the northern and the western states in the country. Implementation of this scheme will, however, depend on the co-operation of neighbouring countries.

The second component is Peninsular Rivers Development which is further divided into four parts. The first part envisages transferring the surplus waters of the Mahanadi and the Godavari to the water-short rivers in the south, namely, Krishna, Pennar and the Cavery. This will enable irrigation in the drought-prone areas of Maharashtra, Karnataka, Andhra Pradesh and Tamil Nadu besides providing benefits of power generation and irrigation to Orissa and Madhya Pradesh. This component is also to build dams in Kerala and divert to the east for irrigation of drought-prone areas of Tamil-Nadu apart from providing additional irrigation and water supplies in Kerala. The third part envisages constructing storages and inter-linking small rivers flowing along the West Coast north of south of Tapti. This will not only help in meeting the drinking and industrial needs of Metropolitan Bombay but also in providing irrigation facilities to the coastal areas in Maharashtra and drought-prone areas in Suarashtra and Kutch region of Gujarat. The fourth part envisages inter-linking of southern tributaries of the Yamuna like the Ken and Chambal in addition to construction of dams on small intermediate tributaries to provide irrigation benefits in Ujjain-Indore region of Madhya Pradesh as well as upper areas in Rajasthan, besides benefitting the area in Ken, Dhasan, Betwa and other basins in Uttar Pradesh and Madhya Pradesh.

Full details of the scheme have not yet been published. It should be possible to obtain a good share of water from the Himalayan River Development and also from the proposed interlinking of southern tributaries of the Yamuna.

The scheme in general was discussed at a high level meeting at New Delhi. It was agreed that the state governments will provide outline plans indicating the prospective water requirements in their territories by September 1980. It is the appropriate time for Rajasthan to assess its requirements of water to develop the desert, showing the lifts involved in diverting the water from the proposed National Water Grid.

11

Delivery System for Rural Development: Watershed Development and the Creation of a Watershed Department in Rajasthan

ANIRUDH KRISHNA

It has been apprehended that in the next few decades, India is going to face an enormous challenge in order to meet its basic requirements of biomass, expressed in terms of food, fodder, fuelwood, timber and industrial raw material. There is already a sharp imbalance between requirement and supply of fuelwood and fodder.[1] In the crop production sector, although a tenuous self-sufficiency has been achieved, growth rates of production have been barely keeping pace with the growth of population. In addition, diminishing returns to inputs are being observed in the forty or so districts which constitute the core of the "green revolution" and negative growth rates in several other areas which have not been benefited by this strategy. The main criticism of the strategy of agricultural growth followed in the past is that the strategy has concentrated on a small number of irrigated areas to the exclusion of the vast dryland areas of the country. The position in Rajasthan is similar to that prevailing in the country. If anything, the position is worse because of recurrent droughts caused by low and irregular rainfall. The watershed development programme of Rajasthan has made a successful beginning in tackling some of these issues. The initial field results are encouraging.

Scheme for Dryland Areas

Of the 34 million hectares land area of Rajasthan, 13.3 million hectares is put of cultivation, 2.27 million hectares is forest land, and 13.8 million hectares is waste land. Only 3.3 million hectares is irrigated, of which two-third is irrigated by wells (which are ultimately dependent on rainwater for recharge). Thus, except for about 1.2 million hectares, i.e., roughly 3.5 per cent of land, the remaining 96.5 per cent of Rajasthan's land area is entirely dependent on rainfall for all its requirements of moisture. The conservation and harvesting of rainwater, therefore, assumes a critical significance for the state. On the other hand, only 11 per cent of the precipitation annually available to Rajasthan is harvested. The balance is wasted and flows out leaving heavy erosion in its wake. The average loss of top soil per annum is colossal 18 tonnes per hectare.

Investment in agriculture sector has hitherto been concentrated in irrigated areas. For the country as a whole, two-thirds of the total investment during the planning period has been concentrated on one-third of the area. This skewness in the distribution of government expenditure is reflected in the growth rates of the production. It is estimated that between 1970 to 1973 and 1980 to 1983, fourteen of the 450 districts in the country contributed 25 per cent of the total increase in agriculture production. Simultaneously, a marked decline in the marginal productivity of inputs is being observed in the irrigated areas where the growth rates have been high in the past.

In this background, the approach paper to the Eighth Five Year Plan (1990-95) resolved to "end the neglect of the vast rainfed and dryland areas". The National Watershed Development Project for Rainfed Areas (NWDPRA) was introduced as a fully centrally sponsored scheme (75% grant 25% loan). One watershed of size 1,500 to 2,000 hectares was to be selected in each panchayat samiti having less than 30 per cent irrigated area. A sum of Rs. 136 crores was made available to Rajasthan over the five year period. Simultaneously, a Rs. 74 crores Integrated Watershed Development Project was approved for funding by the World Bank. This is in operation in the districts of Ajmer, Bhilwara, Udaipur and Jodhpur. One watershed of size 25,000 to 30,000 hectares has been selected in each of these districts. These schemes became operational with the formation of the new Watershed Development and Soil Conservation Department in January 1991.

What is a Watershed?

Before we go on to explain the working of these programmes, it would be necessary at this stage to understand what a watershed is and what is the importance of watershed based development in the context of rainfed areas. A watershed is a hydrological unit which is simply explained as a piece of land and which drains at a common point. There is, therefore, a ridge from which water flows down constantly until it reaches the valley where the common point is situated. The idea is to prevent the flow of this water so as to achieve in *situ* soil and moisture conservation. On the basis of this conserved soil and water, production systems based on efficient land use management are introduced. The activities of conservation consist of contour barriers (mainly vegetative), contour plowing, drainage line treatment with low cost structures, water harvesting, etc. The production measures consist of dryland farming systems, silvi pasture, afforestation, farm forestry and horticulture. The choice of production system depends on land tenure and land classification.

The basic idea in each watershed is to create demonstrations of replicable appropriate technology. This implies not only that the technology results in a higher level of production on a sustainable base but that it is both cheap and relies on local materials. One of the constraints which we have put upon ourselves for this reason is that the investment in conservation should not exceed 8 to 10 mandays per *bigha*. In response to the challenge to evolve new technologies, work methods and departmental culture, the department has introduced some innovations. There are two main strengths of the approach being followed by the project authorities:

(i) The technology used is low-cost, relies on local raw material, and in many instances, has been evolved by the beneficiaries themselves.

(ii) There is very active involvement of the people residing in the watershed. Several mechanisms to ensure active people's participation have been built into the standard operating procedures.

The New Department

In January, 1991, the government set up a separate Watershed Development and Soil Conservation Department[2] with the objective of implementing the National Watershed Development Programme for Rainfed Areas (NWDPRA) and the Integrated Watershed Development Programme (IWDP). The department was to have its own field staff drawn from the different line departments and was to evolve its own procedures for meeting the objective of integrated area development. A policy decision was subsequently taken in June, 1991 that all soil conservation works being undertaken in the state under any scheme of funding shall comprise watershed development works only. Thus, under Desert Development Programme (DDP), Drought-Prone Area Programme (DPAP), and Jawahar Rozgar Yojna (JRY) etc., only integrated watershed development works are being taken up. The difference with the past is that not only are soil and water conservation treatments being installed, but improved production systems are also introduced in the areas so treated.

The department has taken up work on 3.5 lakh hectares of land under NWDPRA, 1.0 lakh hectares under IWDP and about 1.0 lakh hectares under the other rural development programme. During its first year of existence, i.e., 1991-92, the department achieved a financial outlay of Rs. 2,188 lakhs and 53,856 hectares of land was treated under watershed development works. During the first half of 1992-93, financial outlay has been Rs. 1,409 lakhs and treatment has commenced on 72,656 hectares of land; 39.37 lakh saplings were planted and 10,998 hectares were developed as pasture in the watershed areas in this six months period.

The department has a strength of 1,875 persons. It is organised into sixty-two field units, seventeen divisions, three circles and two zones (the subject is dealt with in more detail ahead). The World Bank project—IWDP—has a separate organisation headed by a project coordinator of the rank of additional director, with dedicated project staff at the field level. At each level, the character of the staff is multidisciplinary. They are drawn from the disciplines of agriculture, soil conservation, forestry and horticulture.[3]

Method of Work

Selection of Watershed

Watershed atlases have been prepared for each district of the state indicating each watershed of size 1,500 to 2,000 hectares, approximately. This identification has been done on the basis of remote-sensing data obtained from NRSA and aerial photos available to the State Remote Sensing Agency, Jodhpur. Further, watersheds have been prioritised in terms of their susceptibility to erosion. The department selects watersheds for treatment in this order of priority.

Watershed Treatments

The typical sorts of treatment in a watershed include all arable and non-arable lands and drainage lines situated within the boundaries of the watershed. The idea is to conserve soil and rainwater and introduce scientific land use systems with the objective of enhancing the production of food, fodder, and fuelwood on a sustainable basis. Unit costs are kept low with the intention that the treatments should be replicable in the areas which cannot be treated with government support within a reasonable period of time. Thus, costs on conservation measures are kept within a limit of Rs. 1,000 per hectare which translates to 10 to 12 mandays per *bigha*. Furthermore, only local materials are used, especially vegetative material. The overhauling of methods and technology has made the department take a close look at indigenous technology. Especially when dealing with vegetative conservation methods, which cannot but be area-specific, there has been much which has been adapted from local wisdom.

Project Objectives

The basic objective of the project is to create replicable models for the integrated development of dryland areas. This would *inter alia* include:

- To conserve, upgrade and utilise natural resource endowments to ensure perpetual availability of food, fodder, fibre, timber and biomass in a sustainable manner.
- Generation of employment in rural sector to create permanent assets during the project period. To enhance employment opportunities in the post-project period on a continuing basis.
- Improvement of ecological balance and production environment through scientific management of land and rainwater. This would be done through *insitu* moisture conservation, i.e., runoff management and scientific production systems.

- Reduction of inequalities between irrigated and rainfed areas. Improved cash flows for residents of rainfed areas.

Thus, each selected watershed is treated for a period of five to seven years. The department has an opportunity to observe the results of its works from year to year and make corrections as necessary.

People's Participation

Private Lands

Work on private lands is executed by the landowner himself. In addition, he has to contribute 10 per cent of the total cost either in cash or by way of labour. There is thus a very active discussion of the cost and benefit of the work before the farmer agrees to do it. It is our belief that if a farmer takes up the work in his land and also pays a part of the cost, he will do so because he has positive expectations from the treatments; and not merely as a way to get labour income from a government scheme (even though the importance of wage income cannot be minimised). The contribution of 10 per cent is no doubt token, but it is one way to ensure that the farmer has a personal stake in the work.[4]

Users' Committees

On common lands, i.e., revenue, forest and panchayat land situated within the watershed, work is executed through a committee constituted by the residents of the watershed by elections from amongst themselves. Special provisions have been made in the delegation of financial powers to allow delegation of powers to the users' committees. This is another attempt to foster a feeling of ownership of the work in the beneficiary group. Execution of work on common lands through the departmental machinery often results in a feeling of indifference among the local population due to their exclusion from the work. This has resulted in many treated pastures and village forests being destroyed, either deliberately or through neglect, once the government machinery has stopped physically protecting the area. In the Watershed Development Programme, the planning, implementation and maintenance of works on common lands are done through users', committees. These are formed out of the traditional users of the village common land. The technical supervision is provided by the department. However, the planning of work, choice of species, protection and sharing of usufruct are the functions of the committee.

The users' committees are being made partners in joint forest management and are also being registered as co-operative societies. In nearly two years of their existence, these committees have played a very useful role in mobilisation of village opinion and generating a fairly high level of enthusiasm, unusual for a government programme. "The obvious and immediate benefits to the farmers of Cheetakhera and Andheri Deori[5] has made them more willing partners in the concept of watershed treatment" (*Economic Times*, New Delhi, October 19-20, 1992).

Mitra Kisan and Gopal

Four to five persons per village are selected and intensively trained in the new technique of watershed development. Training sessions are held three times a year. In addition, one postgraduate apprentice (M.Sc. Agri.) is also deployed in each watershed. By these methods, an effort is made to transfer the latest scientific and research findings to the residents of watershed. Simultaneously, the apprentice gains an appreciation of the field reality.

Participative Planning

Video films and extensive discussions were used in the initial phase to familiarise people about the kind of work to be done in the watershed development and the benefits thereof. The implementation methodology in each watershed was decided during these discussions. The usual method of work consists of a three-step procedure:

(i) The watershed area is surveyed extensively. The surveys done include topographical survey, soil classification, resource survey and household survey.

(ii) A scientific land use plan is prepared on the basis of the results of these surveys. A watershed treatment plan is also prepared which indicates the treatments required to arrive at the indicated future land use.

(iii) The scientific land use and treatment plans are then presented in a meeting with the watershed area residents and become the basis for discussions. The discussions result in the formation of a final treatment plan, which can correctly be called as the "people's plan". The department's effort is to keep the final plan close to the scientific plan, but this is not always easy.

Emphasis on Training

The staff has had to be trained in the new technologies and work methods. Extensive training sessions for staff of all levels have been conducted with the aid of several institutions of repute including Rajasthan Agriculture University, CAZRI, Forestry Training Institute etc. Members of farmers' and users' committees have also been trained. NGOs are being involved to impart training in the aspects involving people's participation and longer-term project sustainability.

Learning from Other States

Joint groups of officers, field level workers and farmers were sent on study tours to the states of Karnataka, Andhra Pradesh and Maharashtra, which are the pioneers in watershed development. The experience gained during these tours greatly facilitated the understanding of work methods and techniques.

Technical Manual for Watershed Development

Based on the works observed in the first year and the information gathered from other states, a detailed technical manual was prepared and circulated to all officers and staff. This exhaustive manual covers all aspects of conservation, water-harvesting structures, crop husbandry, horticulture, afforestation and nursery raising. The manual is to be revised from year to year. It must be mentioned that the manual is neither exhaustive nor prescriptive. The effort has been to compile and make available a menu of technologies which have been developed so far. Further, innovation is not only allowed, it is encouraged.

Privatisation of Topographical Surveys

In an effort to enchance the capabilities of the department without adding to staff expenditures and at the same time, to provide a source of income to unemployed technical personnel, a decision was taken to contract out these surveys to unemployed engineers at fixed rates per hectares. The decision enabled 165 such persons to earn between Rs. 2,000 to Rs. 4,000 per month in this period.

Concurrent Evaluation

The Rajasthan Agriculture University and the Society for Promotion of Wasteland Development, New Delhi—a well known NGO in the field—have been assigned the tasks of M&E. These third-party evaluations shall be helpful in deciding future strategy.

Research Support

Since the work of watershed development is new and the technology is area-specific, much emphasis has to be put on a close linkage between scientists and field workers. The department has signed memoranda of agreements with Rajasthan Agriculture University, CAZRI and ICRISAT for specific action-oriented research projects.

The work has been proceeding well so far. Since the interventions are biological and agronomic, the ultimate results, however, will only be finally known after three to four years.

Decision Affecting Administrative Procedures

The administrative structure of the department reveals that the head office has the director and supporting staff. Apart from overall charge and responsibility, it has the duties of budget control, audit and accounts (senior accounts officer), administration and establishment (joint director, administration), monitoring and evaluation (deputy director, M&E), training (deputy director, training), and technical supervision (a technical cell, reporting to the additional director, Jaipur zone). The field structure consists of two additional directors (zone), three joint directors (circle), seventeen deputy directors (division) and sixty-two field units. The field units implement all the physical work of the department, the rest of the levels are supervisory.[6]

The Link Officer System

Effective communication between the field and headquarters levels is essential in an evolving department. Policy decisions have to address the field reality and field level personnel have to be kept abreast of the fast changing policy framework. The hierarchical system of government departments often leads to a programme being closely controlled and monitored by the highest levels of government. Often, the bulk of the decisions are also taken at these levels, which are somewhat remote to the operational levels of project implementation. This does help in reducing the initial inertia and ensuring a broad support-base at the policy level. However, the gap between the level which decides the task to be performed and that which performs the task can sometimes result in inflexibility being built into the system. Feedback gets muted and garbled in travelling up the levels of hierarchy between the operational level and those which take the final decisions. Often these inflexibility may result in a departure from practicality and realism at the field level. This would result in a loss of commitment towards the project amongst the project staff and the beneficiary group. This is especially true when a new innovative programme gets launched. The existing system of written communication up and down the hierarchy results in delays and frequent misunderstanding, especially when the department is large.

In order to handle these complexities, a system of link officers was evolved. Six officers of the head office, of a level equivalent to unit officers,[7] have been nominated as link officers. These officers have been given the responsibility for two to three field divisions each, in addition to their regular staff functions. It is the duty of the link officer to ensure that the divisions in their charge receive, understand and implement the directions of the head office and conversely, that head office is quick in responding to the messages from the field. The officers tour for at least eight to nine days every month. The director chairs a monthly meeting which is attended by the officers of head officer and the link officers. The link officers present their reports for the previous month. The system is an informal one of verbal reports and guidance. However, a register has been maintained at each watershed for visitors or inspecting officer to record their observations. At the monthly meeting, the performance of each division is evaluated in terms of quality of work,[8] adherence to the calendar of activities (which is crucial when the field work relies mostly on biological and agronomic interventions) and response to recent policy directions. The link officers are also advised what are the priority items to look out for in the next round of visits.

In many instances, policy decisions have been reviewed on the report of the link officer. This system does not absolve the senior officers of head office and the circles of their touring duties. It ensures a reliable and continuous linkage between the fields and head office especially during periods when senior officers get tied down by staff responsibility.

State Level Policy and Implementation Committee

The nature of work involved in watershed development makes decentralised decision-making imperative. However, the dangers inherent in decentralisation become apparent when the commitment of the policy level towards the project becomes uncertain. It, therefore, becomes necessary to build structures which maintain operational flexibility while having a strong policy level support which is necessary to ensure a smooth flow of funds, deployment and continuity of suitable personnel in the project and the ability to withstand early criticisms and complaints. Such policy level support has been attempted through setting up high level policy and supervision committees to guide and support the project staff. These committees are charged with the responsibility of deciding on questions of policy as they have emerged in the course of implementation. The support of the committee was crucial in the early stages, when firm direction has to be given to the project, e.g., the decisions for employment of users' committees, privatisation of surveys and involvement of NGOs were novel for a government department and would not have been possible without the support of the high level committee. Similarly, problems of inter-department co-ordination are important in a work which is multidisciplinary by nature. The committee has enabled a smoother resolution of these issues. As soon as the major policy issues are identified and resolved, the project staff is empowered by these decisions and attains the degree of authority necessary for decentralised decision-making.

Decision for a Divisionalised Departmental Structure

When it was first created, the department consisted of about sixty field units. All finances for works were expended by these units and budgets for the same were passed on directly from the head office to the field units. The field units were themselves responsible for ensuring that they fulfilled the annual targets allotted to them. Officers of the level of the division and the circle existed but they had a sort of overall, ill-defined mandate for supervision over the field units, without any specific original responsibility. Thus, on the one hand, there were a large number of units which were being virtually controlled from the head office, and on the other hand, offices did exist which were supposed to act as a channel between the head office and the field but could always claim that they were rendered ineffective because of the absence of any formal financial or administrative powers. The situation needed to be remedied by first identifying which was the crucial layer of the pyramid which could act as a nodal point.

The factors affecting this decision included the span of control of the levels of zone, circle and division; their relative effectiveness in controlling field performance; the quality of officer presently available at each of these levels and the ease with which the field officer could contact the nodal supervisory officer. After analysis and discussions with each of the levels in the hierarchy, it was decided that strengthening the divisional officer would be most effective. In mid-1992, the orders empowering the divisions with financial and administrative authority were issued. Funds for works are now sanctioned to the divisions and the divisions make payments to the units on the basis of the work accounts rendered by them (i.e., from the unit to the division). All correspondence to the head office regarding budget and progress reporting is done by the divisional officer. The divisional officer has also been given some powers of an establishment nature in respect of the staff posted in the units in his jurisdiction. Responsibility for performance has now been transferred from the unit to the division.

This change has been welcomed by both the unit and the divisional levels. The units now feel free to devote their time to watershed development without being impeded by financial and administrative paper work. The divisional officers have been given the powers that they felt had been denied to them earlier. There has been a relatively smooth transference from the old system to the new. In part this has been such because the change was preceded by several rounds of discussions at departmental workshops. When the change came, it had been anticipated for some time by all. However, in order to react quickly to any unforeseen problems in the field, the link officers were instructed to specially report on the manner of transition during the two months following the announcement of the new dispensation. In addition, since the new dispensation was accompanied by a new set of financial procedures, six officers of the accounts section of the head office were also temporarily sent out to specified divisions to smooth enthings out. The issue of smooth transition was on the agenda of every departmental conference held between March and November, 1992. The new administrative structure is now fully operational.

NOTES

1. The Forest Survey (1987) has estimated that against a requirement of 235 million cum. of fuelwood, the supply is only 40 million cum. The supply of dry fodder is only 450 million tonnes against a requirement of 932 million tonnes and the supply of industrial timber is 12 million cum. against a requirement of 27 million cum.
2. There was earlier a Soil Conservation Wing functioning under the agriculture directorate. This was subsumed under the new organisation.
3. The work which has been done in Rajasthan in this short period of almost two years has been widely appreciated. In respect of both the IWDP and the NWDPRA, we are far ahead of the work done in other states. Rajasthan has been selected by the GoI to host the National Seminar on Watershed Development in February, 1993. The World Bank had this to say: "The Rajasthan component of the project is making good progress.... The project has done well on the people's issue and has been able to deploy competent and enthusiastic staff."
4. As the benefit of the new technology becomes more visible, it is proposed to enhance this percentage. Eventually, if a significant part of the area has to be covered, the entire figure would have to approach 100.
5. Two of ongoing watersheds in Ajmer district.
6. The World Bank-funded IWDP has a dedicated project implementation unit under the charge of a third additional director who has been nominated by project coordinator, IWDP. In each of the four locations where this project is being implemented, there is separate project staff, under the control of a deputy director, designated as project director. Apart from these changes in nomenclature, the functioning of the IWDP set-up is similar to that in the rest of the department.
7. This level of officer cannot ride roughshod over the divisional officer, but is experienced enough to be able to take decisions of the spot.
8. Quantum of work is already ascertained through the monthly progress reports.

12

Community-Based Organisations and Panchayats as Instruments of Governance in the Sphere of Watershed Development

RAKESH HOOJA AND RAKSHAT HOOJA

The Government of India has long considered watershed development as a catalyst for development activities in rainfed areas. Accordingly, during the last twenty-five years, most programmes of the Ministries of Rural Development, Agriculture and Environment have followed the watershed development approach. During this period, a number of non-governmental organisations (NGOs) and panchayat bodies/District Rural Development Agencies (DRDAs) as well as state governments and grass roots researchers gained considerable experience in the implementation of watershed projects. Based on their feedback and the report of the Hanumantha Rao Committee on Drought Prone Areas Programme (DPAP) and Desert Development Programme (DDP), the Government of India had, in 1994, developed 'Guidelines for Implementation of Watershed Programmes'. These guidelines were subsequently (following the publication of 'Common Principles for Watershed Development', prepared jointly by the Rainfed Farming Division of the Ministry of Agriculture and the Ministry of Rural Development in 2000) revised in 2001 by the Department of Land Resources of the Ministry of Rural Development and circulated as the Guidelines for Watershed Development (Revised 2001) (popularly referred to as Common Guidelines or GWD).

The GWD aimed to bring people to the centre stage and push the administration towards a facilitating and assisting role. The focus was on the enhancement of the viability and quality of rural livelihood support systems as indicated by the objectives of watershed development mentioned in the Common Guidelines, which referred to developing wastelands/degraded lands/drought prone and desert areas on watershed basis; and also called for attention to improving the socio-economic condition of the resource-poor sections inhabiting the programme areas, employment generation, poverty alleviation, community empowerment, as well as encouraging the village community to make use of simple and affordable technological solutions and institutional arrangements.

The Panchayat Raj Institutions (PRIs), which are India's constitutionally recognised local government bodies, had an important role in Watershed Development Programmes (WDPs) as per the Common Guidelines, having the right to monitor and review the implementation of the programmes and provide guidance. At the village level, the Gram Panchayat was to be involved in the implementation of the programme. The Zila Parishad, Panchayat Samiti and the Gram Panchayat could also take on the responsibility of implementing a cluster of watershed projects in the capacity of a Project Implementation Agency (PIA). However, for this, they had to compete with other NGOs, government organisations (GOs), community-based organisations (CBOs) or other support organisations to get selected as a PIA.

The Common Guidelines proposed a user-friendly organisational structure with Watershed Associations (WAs) and elected/nominated Watershed Committees (WCs) undertaking the field/village level implementation of each watershed under the supervision of the PIA (which was to manage 10-12 watersheds), and with technical inputs from experts who comprised the Watershed Development Team (WDT).

A district watershed committee with multi-source membership helped ensure coordination as well as the selection of PIAs and members of WDTs and in the approval of detailed action plans for watershed projects in the district.

To recapitulate, the main institutional and administrative arrangements as per Common Guidelines of 2001. can be summarised as follows:

1. A State Watershed Development Committee (SWDC) constituted to ensure coordination among various government departments/institutions and voluntary agencies. This committee was to meet twice a year to monitor, review and evaluate the progress of implementation of WDPs.
2. The WDPs to be implemented through the Zila Parishads/ DRDAs.
3. To ensure coordination at the district level, a District Watershed Development Committee (DWDC), headed by the Chairman, Zila Parishad, or DRDA, constituted with a multi-source membership of officials and non-officials. The role of DWDC was to advise and assist the Zila Parishad/DRDA on matters regarding selection of PIAs, and of members of WDTs, as well as training, activities of community organisation etc. The committee was also to approve the detailed action plan for watershed development projects in the district.
4. The PRIs had the right to monitor and review the implementation of the WDPs and provide guidance for improvements in the administrative arrangements and procedures with a view to ensuring convergence with other programmes of the Ministry of Rural Development. At the village level, the Gram Panchayat was to be involved especially in community organisation activities and training. The Zila Parishads, Panchayat Samitis and the Gram Panchayat could also take on the responsibility of implementing a cluster of watershed projects as a PIA. However, for this, they had to follow norms prescribed for PIAs under the Common Guidelines, and had to compete with other NGOs, GOs, or CBOs or other support organisations to get selected as a PIA.
5. The projects were implemented at the field level by the WCs under the overall supervision and guidance of PIAs. PIAs to normally be assigned 10-12 watershed projects covering an area ranging between 5,000-6,000 hectares and to be responsible for constituting four-member WDT for planning, implementing and

reviewing of watershed programmes through village level organisations, and also maintenance of accounts. The members of PIAs to preferably be selected from amongst PRIs, government departments or a reputed NGO, which had been active in the field of watershed development or similar area of development.

6. The WDT had a minimum of four members, one each from the disciplines of forestry/plant science, animal sciences, civil/agricultural engineering and social sciences.
7. The Common Guidelines envisaged the following village level institutions to enable the communities to take up participatory watershed management:
   a. *Self-Help Groups* (SHGs): Constituted in the watershed area with the help of the WDT, these were to be homogenous groups, who were dependent on the watershed but not owning land in the watershed areas, such as agricultural labourers, shepherds etc.
   b. *User Groups (UGs)*: These groups to consist of persons most likely to derive direct benefits from particular watershed work and holding land within the watershed areas. The UGs were to eventually take over the operation and maintenance of completed community works.
   c. *Watershed Associations* (WAs): Members of the community directly or indirectly dependent on a watershed area organised into a WA, which should be a registered society as per Registration of Societies Act, 1860. The WA to meet at least twice a year to improve, monitor and review the watershed development plan, nominate members of the watershed committee etc. Wherever a watershed was coterminous with a village panchayat, the Gram Sabha of the Gram Panchayat concerned could be designated as the WA.
   d. *Watershed Committees* (WCs): The WC to consist of members nominated by WA from amongst the UGs, SHGs, Gram Panchayat and a member of the WDT for carrying out the day-to-day activities of the Watershed Development Project. Each WC had a secretary to be a paid on an honorarium basis by the WA.

Just as the initial projects under the Common Guidelines were beginning to bear fruit, with growing awareness about the procedures, schemes, benefits etc. among the beneficiaries and the potential PIAs, as well as CBOs, NGOs, local government staff etc., the government (Ministry of Rural Development) decided to launch a new 'Haryali' scheme in January 2003 under which it replaced the Common Guidelines with the Haryali Guidelines. The Haryali Guidelines stipulated that CBOs, comprising UGs, SHGs, WCs and even the WAs, as well as NGOs, which have often proved to be efficient PIAs, would cease to have any role in watershed development and that, henceforth, the PRIs (Gram Sabhas, Gram Panchayats, Block Panchayat Samitis and Zila Parishads) are to manage and implement watershed development activities.

Some changes brought about by the Haryali Guidelines are:

- The Haryali Guidelines have done away with WAs and WCs; their functions being transferred to the Gram Panchayats. The Gram Panchayat is now expected to perform both executive functions (even being responsible for day-to-day activities) as also governance functions. The UGs and SHGs are now to be constituted by the PRIs and the role of the UG has been reduced to mere maintenance of the created assets while its earlier role in planning and executing development works has been ended.
- Under the 2001 Guidelines a large variety of organisations were allowed to function as PIAs, but preference was given to PRIs, a line department of government, or a reputed NGO/other corporate body, in that order. The Haryali Guidelines, however, indicate that an intermediate panchayat (Block Panchayat/ Panchayat Samiti/Mandal Panchayat/Taluka Panchayat) may be the PIA for all the projects sanctioned in a block/taluka/tehsil. In case the intermediate panchayat is not adequately empowered, then the Zila Parishad either may itself act as PIA or appoint a suitable line department (agriculture, social forestry/soil conservation etc.) or an agency of state government/university/institute as PIA, failing which a reputed NGO with experience of having worked as a PIA could be considered. Thus, eventually, it is only the intermediate panchayat bodies, which so far were not associated with watershed development, who are to be the PIAs.
- The District Watershed Committee has been dispensed with, through one wonders why the Haryali Guidelines have chosen to deprive the Zila Parishad of the advice/guidance of a district level body of social and technical experts and representatives of

various stakeholders (GOs, NGOs, CBOs and other support organisations) intimately connected with watershed activities.

- The 2001 Guidelines had provided for a watershed secretary for each watershed to be locally engaged on honorarium basis by the WA to work under the supervision of the chairman of the WC. As per the Haryali Guidelines, the Secretary, Gram Panchayat (a government employee) is to assist the Chairman, Gram Panchayat.
- Haryali has left state governments wondering whether their watershed directorates should continue to be attached to the agriculture department and to provide technical/techno-economic inputs, or whether the watershed directorates should be transferred to the state Panchayat department where it would emphasise on participatory and socio-political aspects of watersheds.
- While there is considerable commonality about the objectives of watershed development, the two guidelines differ somewhat wherein the 2001 Guidelines stress developing wastelands/degraded lands, drought prone and desert areas keeping in view capability of the land and local needs and conditions, while the Haryali Guidelines talk of harnessing every drop of rainwater for irrigation, horticulture, floriculture, pasture development/fisheries etc. to create sustainable sources of income for the village community, and provide drinking water supplies. However, where the two guidelines totally part company is that the Haryali Guidelines talk about ensuring overall development of rural areas though GPs and creating regular sources of income for the panchayats as a major objective. The earlier guidelines were silent regarding any such objective.

Evidently, the purpose behind introducing the Haryali Guidelines was to strengthen the PRIs. There exists a group of activists and scholars who feel that such 'parallel' bodies like WAs should either not be permitted to exist or should function as subordinate entities of the PRIs who they see as the third tier of Indian federalism. They often advocate that to strengthen PRIs, it is necessary to abolish the special watershed-related bodies.

The present authors tend to disagree with the view that it is necessary that only panchayat bodies represent the local community or act on its behalf. Not every member of the Gram Sabha is a user or beneficiary, while in the case of a watershed user group every member is user and beneficiary as well as an implementer-cum-manager. In fact, because of their nature, traditions, functions, size and frequency of meetings etc., Gram Panchayats function in a manner very different from watershed user groups and have different sorts of relationships with their members. The Gram Panchayat leadership may also have a wide band of concerns and interests of which watersheds may form a miniscule or peripheral part, while for the watershed user group the watershed is a major part of their existence. It is thus necessary for panchayat bodies and local user groups to not only co-exist but also to supplement, complement and strengthen each other. Perhaps, a small amount of creative tension between PRIs and user groups, or other CBOs, may even prove to be good for the community and for the watershed development efforts.

It needs to be questioned whether using of Haryali as a tool for empowering PRIs, to the extent of displacing the WCs, WAs, and other user groups and SHGs, is the appropriate or best way to empower PRIs. Without the devolution or decentralisation of appropriate powers and authority from the state governments to the various levels of PRIs, as is called for in the 73rd Constitution Amendment Act, it is difficult to expect real empowerment of PRIs. It is such devolution of powers (which has still to be effected properly in most states), rather than trying to take over functions from the primary community-based stakeholder groups (UGs/SHGs), which would really empower panchayat bodies.

It may also be pointed out that experience in the field has shown that Gram Sabhas have not been able to perform the required functions effectively due to their large size, unorganised membership and high degree of heterogeneity.

A Gram Panchayat may comprise one, two or three villages. Each village and its attendant hamlets may normally have a population of between 1,500 and 3,000 persons (though the population of a particular village may go up to even 15,000). The Gram Sabha's/Gram Panchayat's average population in most states would be between 2,000 and 6,000 (though in one state the average is about 21,000). Thus, the Gram Panchayat cannot be considered to be grassroots in today's India; instead, it would be a hamlet, or a ward, or a single watershed/group of 2-3 contiguous watersheds, which would be the real grassroots.

Also, the Gram Panchayats are territorial units and not ecological units. Thus, a watershed may cut across boundaries of Gram Panchayats or a Gram Panchayat may cut across watershed boundaries. Where the watershed area is coterminous with the Gram Panchayat, the Gram Sabha could function as the WA. However, if there are several watersheds within the area of a Gram Panchayat, then each such watershed area, and the people depending upon the resources of the watershed, should have their own WA.

There is need for an executive body elected from the UGs and SHGs to manage the watershed programme on a day-to-day basis. The Gram Panchayat cannot entirely replace the WC in its roles and responsibilities as it takes away decision-making from the direct stakeholders. This is particularly so if the watershed area is only a part of the entire village. Of course, if the Sarpanch of the Gram Panchayat is a member of a UG/SHG, then he/she could also be elected as chairperson of the WC. One option could be that the WC could also be considered as a standing committee of the Gram Panchayat for purposes of managing the watershed programme.

It is unfortunate that the Haryali Guidelines have reduced the percentage of budgets available for community organisation and for training and capacity building.

We, as informed observers, are clear in our mind that the prime objective of all watershed efforts has to be improved land-water management and rural livelihoods including the generation of incomes and optimal use of natural resources, and not the strengthening of any governmental institution like the panchayats, for which other steps could be undertaken.

Similarly, it does not make sense to let the capabilities for watershed work that CBOs, NGOs, GOs, and other agencies working as PIAs have built up over time go waste by divorcing them from watershed activities. Since both PRIs and CBOs lack in technical expertise, the role of district and lower level staff of state government departments and of experienced NGOs having professional staff in providing technical inputs either as PIAs or through other coordinating mechanisms, should not be underemphasised.

Thus, all things considered, it appears that the Haryali Guidelines would have done well to have retained provisions for WCs, WAs, and effective watershed users groups as well as NGOs etc. working in coordination with panchayat bodies. This would not only enable the achievement of natural resource related land-water management goals, but also result in the greater importance of panchayat bodies. Accordingly, appropriate mechanisms and procedures for promoting harmonious relations between participatory watershed groups (which are community-based organisations) and Gram Panchayats need to be innovated with and put into practice.

As stated at the outset, PRIs constitute the third tier of Indian federalism. Watershed bodies working in parallel to, or at levels lower than, panchayat bodies comprise (in a manner similar to the case of water users associations in the irrigation sector) part of a fourth tier whose importance is gradually being understood. Neither needs to replace the other, both can co-exist and reinforce each other, as also coordinate and interact with functionaries of the state government (second tier of Indian federalism) for performing the complex tasks of watershed development and management.

REFERENCES

Chandrudu, M.V. Rama, 2003 (a). "Historical Transact of Watershed Policies in India: Shifts in Content and Philosophy and Their Implications". Paper circulated at Orientation and Training Workshop on Watershed Development and Management, organised by Indian Society of Ecological Economics and Gujarat Institute of Development Research at Ahmedabad from 16 to 19 October 2003.

—, 2003 (b). "Haryali and Other Guidelines: Issues, Concerns and Gaps" in *Haryali Guidelines: Issues and Concerns,* WASSAN and APRLP, Hyderabad.

Hooja, Meenakshi and Rakesh Hooja, 2001. "Institutions of Self-Government or Agencies for Development: Panchayats in India", *South Asian Journal of Socio-Politics Studies.* Vol. 2, No.1, January-June 2001.

Hooja, Rakesh, 1985. "People's Participation in Comprehensive Watershed Area Development" in Hoshiar Singh (ed.), *Rural Development in India,* Printwell, Jaipur; reprinted in Rakesh Hooja, *Administrative Interventions in Rural Development,* Rawat Publications, Jaipur, 1987.

—, 2003. "Below the Third Tier: Water Users Associations In India" in on-line research library on federalism of Forum of Federations—an international network on federalism based in Ottawa Canada (www.forumfed.org), revised version published in *The Indian Journal of Federal Studies*. No.1, March 2004.

—, 2004. "Water Users Associations as the Fourth Tier of Indian Federalism", *South Asian Journal of Socio-Political Studies*, Vol. 4. No.2, June 2004.

— 2006. *The Management of Water* for *Agriculture: Essays in Management of Irrigation, Watersheds and Drainage*, Rawat Publications, Jaipur.

— and Ganesh Pangare, 2002. "Some Reflections on Software Aspects of Water Policy: CAD, PIM, Watershed Development and Management and Water Harvesting", Paper presented at a Round Table in Water Policy organized by INSEE at New Delhi, May 10, 2002. Its revised version printed as India NPIM Working Paper No. 13, New Delhi, January 2003.

— and Rakshat Hooja, 2003. "The Changing Role of Panchayat Bodies in Watershed Development". Paper presented at National Seminar on Panchayats at Madhya Pradesh Institute of Social Science Research, Ujjain, in December 2003 and published in the seminar volume *Functioning of Panchayat Raj System, 2005* edited by Y.K. Sisodia, Rawat, Jaipur. (This paper was also invited for presentation at the IWMI-Tata Partnership Annual Researchers meet on Water at Anand from 17 to 19 February 2004).

Hooja, Rakshat, 2004. "Why Only Panchayats? Other Local Associations Can Also Manage Watersheds Effectively", *Down to Earth*, Vol. 12, No. 18, February 15, 2004.

Indian Network on Participatory Irrigation Management, 2002. Land-Water Management and Johads: Process Documentation of the Activities and Approach of Tarun Bharat Sangh—an unpublished report by Ganesh Pangare, Talha Jamal and Rakshat Hooja written for Ford Foundation.

Jamal, Talha, Rakshat Hooja and Ganesh Pangare, 2008. "Johad Versus Watershed: A Case Study of Tarun Bharat Sangh's Approach to Water Harvesting and Conservation for Drought Mitigation" in Jasveen Jairath, Vishva Ballabh, and Khalid Hussain (eds.), *Droughts and Integrated Water Resources Management in South Asia: Issues, Alternatives and Futures* (publication sponsored by SaciWATERs Hyderabad), Sage India, New Delhi.

Ministry of Agriculture, Government of India (2000). *Guidelines for National Watershed Development for Rainfed Areas*, New Delhi.

Ministry of Rural Development, Government of India (1994). *Guidelines for Watershed Development*. New Delhi.

Ministry of Rural Development, Government of India (2001). *Guidelines for Watershed Development (Revised 2001)*, New Delhi.

Ministry of Rural Development, Government of India (2003). *Haryali*. New Delhi.

OIKOS, 2000. *Social and Institutional Issues in Watershed Management in India.* Oikos, New Ddhi and IIRR Phillipines.

WASSAN and APRLP. 2003. *Haryali Guidelines: Issues and Concerns, Suggestions and Recommendations of Consultations and National Workshop*, Hyderabad.

13

Watershed Plus in Rajasthan

M.K. KHANNA

Rajasthan is the largest state in terms of area in India, with 10 per cent of the land of the country, but it has only one per cent of the water resources. Agriculture is still the primary source of incomes and livelihoods in the state. Given this situation, no wonder that productivity of most crops is well below the national level. About 34 per cent of the total land area of the state can be categorised as wastelands. Over the years, most of the resources have gone into development of irrigation, mostly canal as well as exploitation of the underground water. Groundwater sources are limited and have reached the level of saturation of potential, with 81 blocks out of 237 showing overdrawal of water. If India is to achieve a Second Green Revolution, as is now being talked about in the press, there is need to improve the productivity of the wastelands for unirrigated crops and allied activities. The watershed programme is one such effort, which, if properly implemented, can deliver results in the more deprived areas. Experience has shown that the watershed approach has a lot of potential and can yield results provided some emerging issues are dealt with effectively.

What is a Watershed?

A watershed is a natural hydrological entity; a land surface, from which the run-off water flows into a definable surface drain, stream or river having a single outlet. It is an ideal unit for preparation of integrated area development plans for checking land degradation, water and soil losses and thereby improving the environment. The term 'watershed' is synonymous to catchment, sub-catchment, sub-watershed, macro-watershed, micro-watershed, etc. In Rajasthan, macro-watersheds are classified as covering an area from 1,000 to 10,000 hectares while micro-watershed as covering an area from 100 to 1,000 hectares. The general components of watershed development are:

- Community mobilisation and capacity building.
- Human resource development and training.
- Institutional development.
- Technical aspects—soil, water and land management (conservation and use).
- Moisture conservation.
- Afforestation and pasture development.
- Agricultural development—diversification and productivity enhancement.
- Livestock management.
- Rural energy management.
- Procedural simplification.
- Reformed agriculture extension and dissemination of improved practices.

The Rajasthan Scenario

Rajasthan, with a geographical area of 3,42,239 square kilometers, is now the largest state in India in terms of size. It has a population of 56 million (2001 census), with much denser human settlements in the eastern part, than in the western desert area. More than 60 per cent of the state's area is desert, with sparsely distributed population. The state has wide variations in terrains, social structures, livelihoods and cultural patterns. It has the highest proportion of Scheduled Caste population in the country (over 17%) and a high proportion of Scheduled Tribe population as well (12.4%).

Agriculture and animal husbandry are the principal sources of livelihood in Rajasthan. Although the contribution of other sectors (non-farm and service sectors) to the state has increased, the sectorial portfolio in terms of sources of employment has not diversified significantly.

Agriculture continues to be dependant on rainfall. However, there are sharp regional differences on account of soil quality and terrain, with 23 of the 32 districts averaging over 400 mm annual rainfall. In the region east and south of Aravalli hills, where the land is more fertile, agriculture is relatively more prosperous. On the other hand, nearly 53 per cent of all land in the western desert districts is unfit for farming. Out of the total area of 34.5 million hectares, in 1997-98 only 65 per cent of the land was under cultivation, 14.66 per cent was culturable waste, 10.46 per cent fallow, 7.7 per cent barren, 7.38 per cent under forest, 5.03 per cent under pastures and grazing land, and 4.96 per cent was put to non-agricultural use.

Availability of water has strongly conditioned the nature of agriculture and farming practices in various parts of the state. While Rajasthan has 5 per cent of the country's population and 11 per cent of the country's land, it has only 1 per cent of the country's total water resources. Rainfall in the state varies from a high of 900 mm in the southeast to a low of 190 mm in the western districts. There are frequent failures of the rains, resulting in crop failure and drought. The rainfall is further characterised by frequent dry spell and uneven distribution that seriously affect crop and livestock production.

The landholding pattern in the state is highly skewed. About 50 per cent of the total landholdings are marginal or small, measuring less than 2 hectares in size. The poor generally have marginal quality of land and scarcity of water resources for irrigation is an additional constraint. Only 30 per cent of the total cropped area in the state is covered under irrigation. While 21 per cent of the state's irrigation potential is concentrated in two districts, namely,

Ganganagar and Hanumangarh, in the desert districts of Barmer, Jaisalmer, Churu and Jodhpur, less that 10 per cent of the cropped area is irrigated. As a result, in most parts of southern and western Rajasthan, even with normal rainfall, the household production is barely enough for subsistence.

Agriculture in these areas is also constrained by lack of assured supply of inputs, traditional technologies and cropping systems unsuited to dry land conditions, poor resources and inadequate extension and support services. Research and extension on agriculture in Rajasthan as in the country, have generally focused on more resourceful farmers and better-endowed areas, and largely on irrigated areas and crops. The traditional system of multi-cropped subsistence agriculture practiced by communities in these areas is shifting to commercial mono-cropping, largely because farmers have been led to believe that this is the way of 'prosperity'.

Low productivity of agriculture and the dimension of ecological risk make food security and subsistence the primary concern of farmers. As per the Food Insecurity Atlas of Rural India, Rajasthan, along with Orissa, Chattisgarh and Uttar Pradesh, is among the 'severely insecure' states.

Small and marginal families also join the agricultural wage labour force in large numbers, particularly after crop failures. The availability of work depends on rainfall, and migration of agricultural workers to other districts and states is a common phenomenon. During good rainfall years, families migrate only for the duration of the lean season, but in drought years, the entire family leaves the village for months together.

The watershed development programme was initiated in Rajasthan from 1995-96 under the National Guidelines for Watershed Development. Both the Central Ministry of Rural Development and the state government have provided financial assistance for these projects under DDP (16 districts), DPAP (11 districts), IWDP and erstwhile EAS. Afforestation activities are also taken up adopting a watershed approach. Under combating desertification project (CDP of DDP), sand dune stabilisation, afforestation, shelterbelt plantation, along with farm forestry, are being attempted in the 10 DDP districts through the PRIs. An estimate of the resources available for these projects shows that substantial resources are available in Rajasthan for watershed development. Currently, 5,044 watershed projects are being implemented under various schemes with an outlay of Rs. 1,507 crores. The ongoing watershed development schemes in the state are shown in the Table 1.

In addition to the watershed programme, substantial investment has also been made in water harvesting through other schemes and programmes; for instance, during the famine relief operations in Rajasthan, in 2002-03 and 2003-04, 49,000 water harvesting structures have been upgraded or added. Similarly, under SGRY (special projects) also, such schemes have been taken up in Tonk, Bundi, Jhalawar, Kota and Baran. This is in addition to works on ongoing schemes of the departments of forest, watershed development and PRIs.

Table 1

Rajasthan: Ongoing Watershed Development Schemes

Department	Numbers	Area (in hectares)	Investment (Rs. in lakhs)
Rural Development (DDP, DPAP)	4,226	21,13,000	1,22,185.00
Rural Development (IWDP)	36	2.72,895	14,577.58
Soil Conservation and Watershed Development	1,144	5,48,580	24,789.00
Total	5,406	29,34,475	1,61,551.58

Despite investment of substantial resources in watershed development, there has been little impact in the field, as only 25 lakh out of a total 242 lakh hectares have been treated so far. Nevertheless, in a chronically water-deficit state like Rajasthan, there appears to be little option except to continue with the treatment of watersheds to improve the productivity of land and prevent erosion of valuable top soils. Increase in biomass production is also an important desirable outcome in a predominantly agrarian economy like Rajasthan with 19 per cent of the SGDP based on livestock. For the coming years it is proposed to treat at least 10,00,000 hectares per annum, so as to cover the entire unirrigated agricultural landmass over the coming 25 years. For this purpose, the financial allocations must be increased and the organisational set-up galvanised to meet the challenge.

The qualitative aspect also needs to be kept in mind to get optimum results from the proposed interventions. The experience appears to have been that whereas the physical interventions have been largely successful technically, the other allied activities like afforestation, horticulture, animal husbandry and others have not been as successful. Crop diversification and dissemination of learnings are also not visible. As a result, the impact is very limited. There is also lack of convergence with other rural development schemes and programmes. Even work of replication of improved practices, e.g., water harvesting, has not caught on. Micro-watersheds being developed were meant to be demonstration units to usher in a change in traditional agriculture practices and to contribute to rural transformation in terms of employment and awareness of socio-economic potential for self-development. But, because of departmental nature of the intervention and lack of attention to community mobilisation, the impact has been limited. There is, therefore, need to study the livelihood potential and build backward and forward linkages for non-farm employment, based on the potential of each area. The assistance of organisations like NABARD can be sought for this type of intervention. The whole watershed concept needs to be revisited so as to reorient it by learning lessons from the experience gained in its implementation so far.

It is clear that substantial resources are already committed to watershed development. These can be further augmented by faster implementation of ongoing programmes, tapping other ongoing programmes of the Government of India as well as national and international donor agencies with commensurate benefits to the rural population through added generation of livelihoods and employment, agricultural growth and animal husbandry etc. The wasteland development projects not only assist in upgrading degraded lands but also help in promoting overall development in the socio-economic conditions of the disadvantaged sections of society through social mobilisation, employment generation, additional income generation, drought proofing, poverty alleviation, community empowerment etc.

The success of the programmes in dealing with soil and water conservation however leaves much to be desired. The resources available for watershed development and the potential for such activities in Rajasthan are substantial. With watershed development it is possible to remedy some of the problems identified in implementation of the watershed programme in the state. It can also serve as an ideal means for enhancing the incomes of the beneficiaries and productivity of most land-based occupations.

Some positive progress has been recorded in 2003-04 in speeding up utilisation of funds under DDP, DAP and IWDP with the utilisation topping over Rs. 140 crores in contrast to an earlier average of Rs. 100 crores per annum. The state government has indicated to increase the treated area from a current level of 5 lakh hectares to 10 lakh hectares per annum, with improved quality. Some economists have been talking of need to launch a Second Green Revolution in the country. The watershed programme is an ideal launching pad for such a national effort, as it has the potential to turn the dreams into a reality, if properly implemented.

The watershed approach aims at treating degraded lands with the help of low cost and locally accessed technologies such as *in situ* water conservation, afforestation etc., though participatory approach that seeks to secure close involvement of the user community. The main objectives of the watershed programmes in Rajasthan include:

- Arresting the denudation of land.
- Conservation of water resources by trapping the rain run-off.
- Optimum production of biomass for animal fodder.

Along with:

- Community mobilisation and empowerment.
- More equitable access to livelihood opportunities for landless, marginal farmers and women through self-help groups.
- Enhanced incomes for farmers.
- More opportunities for wage employment for landless.
- Less drudgery for women by adaptation of technology.
- Reduction of vulnerability to drought by more sustainable and productive use of water.
- Reduction of environmental degradation.
- Greater food security for the poor.

- Adoption of an area development approach to harness natural resources and deploy them for benefit of the poor.
- More efficient transfer of technology—utilising technical institutions, district training teams, block training teams and trained volunteers.
- Value addition and support to supplementary livelihoods.
- More opportunities for non-farm sector employment.
- Better access to other government schemes by group formation, awareness and mobilisation.

The people of Rajasthan faced their fourth successive drought in four years in 2003. In the last 50 years, Rajasthan has experienced 45 droughts. In this context, it is important that funds being spent on providing wage employment are utilised gainfully in drought-proofing the state. The watershed programme has shown some promise in this context. However, this programme has lost credibility because of poor execution and leakages. There is thus need to strengthen the planning, implementation and monitoring of the programme, to reap full benefits.

Watershed Programme

Key Concerns

The watershed programme is one of the key strategies that can yield results for sustained poverty alleviation. But the strategy has to be refined to yield better results in terms of poverty alleviation and area development. Some key concerns and remedial actions that need to be highlighted in this context include:

- More efficient and timely pre-project preparation.
- More effective people's participation.
- Familiarisation of field staff with participatory approach.
- Ending insecurity about fund availability at the grassroots level.
- More emphasis on preparatory activities.
- More emphasis on cohesive group formation.
- Transparent criteria for selecting areas and villages to be adopted by the PRIs.
- Developing human resource capabilities.

- Involvement of senior government functionaries and line agencies.
- Better horizontal linkage among various agencies at the district level.
- Exit protocol for withdrawal after project completion.
- Integration of watershed development programmes under common guidelines and cost norms.
- Greater impact through larger allocation of funds for micro-projects.
- Integration of delivery systems of all watershed programmes at the union government level.

At the present pace of sanction it is feared that the total area to be treated will only be covered after many decades, hence there is need to allocate additional resources for watershed programme. This is evident from the following data:

Total geographical area of the state	34.2 lakh ha (100%)
Available area for treatment (other than land put to non-agriculture use, barren land and forest)	29.8 lakh ha (86.95%)
Total area treated or being treated under watershed programme	35.4 lakh ha (11.92% of available area)

Table 2

Rajasthan: Ongoing Watershed Development Schemes

Scheme	Watershed Projects				Total area treated/ under treatment (ha)
	Completed		In Progress		
	No.	Area	No.	Area	
DDP	841	3,15,375	1,737	8,68,500	11,83,875
DPAP	176	66,000	594	2,97,000	3,63,000
IWDP	12	33,986	36	2,72,895	3,06,381
EAS (W/S)	1,123	4,21,125	—	—	4,21,125
NWDPRA	—	9,89,887	1,144	1,30,948	11,20,835
Externally Aided Projects	6	1,52,820	—	—	1,52,820
Total (in lakh hectares)		19.8		15.7	35.48

Learnings from Some Successful Watershed Projects

Some lessons that can be drawn from other successful watershed development programmes, analysed by the Planning Commission, are as follow:

- They devote significant resources to social issues.
- A high proportion of staff members have experience and skills in social mobilisation.
- Project leaders are fully committed to participation and, in most cases, donors or senior officials apply pressure to ensure 'participatory approaches'.
- Project monitoring explicitly checks whether local organisations of users have been formed.
- Staff members have an incentive to undertake participation.
- Communities being organised, have capacity to influence how the field staff works.
- Community is empowered with knowledge and skills to implement the project and value-added activites.

Some Livelihood Support System (LSSs) Adopted in Other States

Some of the LSSs adopted in other states to sustain watershed interventions are enumerated hereunder:

(i)	Crop-based LSSs	• *Dal* processing • Vegetable sun drying • Fruit processing • Coarse cereal processing • Jam, jelly and preservatives • *Achar* and *Murabba*
(ii)	CPR-based	• Charcoal Making • Fibre making • Fish rearing • Agro-forestry • *Ber* cultivation
(iii)	Livestock-based LSSs	• *Khoa* making • Rearing calves • Rearing calves • *Durri* weaving

(iv)	Non-crop-based LSSs	• Vegetables • Flowers cultivation and floriculture • Fruits • Green fodder • Medicinal and aromatic plants
(v)	Support service-based LSSs	• Custom service • In-village seed production • Vermicompost • NPV • Trichogramma • MPTs • Pumpset repair
(vi)	Development of para-extension specialists	• Jankar (GVT approach) • Gopalmitra (BAIF approach)
(vii)	Homestead farming-based LSSs	• Small dairy • Small herds of small ruminants • Fodder banks
(viii)	Nutrition security systems	• Leafy vegetables • Vegetables • Eggs/milk/meat • Fruits • Grain banks

Impact of Micro Watershed in Rajasthan

- There is a positive impact on productivity of various crops.
- Positive impact on the use of agricultural inputs.
- Increase in use of audio-visual aids for promotion of concepts.
- Need for more intensive use of vegetative barriers.
- Need to promote fisheries.
- Need to promote horticulture.
- Need to set up community nurseries at site on farmers' fields.

The basic problem with the implementation of the watershed programmes observed in Rajasthan is one of sustainability. Treated micro watershed fail to sustain the gains in the medium and long terms. There is, therefore, need to create some continuing mechanisms so that the watershed committee and the community sustain their interest in the watersheds after the implementation phase and after the exit protocol is signed and implemented. It may be useful to follow the methodology adopted in some other states where initial investment in watershed development is regarded as the start up of integrated development of the area and the people. Where the watershed project has come to an end, it will be useful to utilise funds from some other schemes to nurture the community groups, particularly groups of BPL and women, by dovetailing resources from other schemes including DPIP.

Suggested Approach

More attention should be devoted to:

- Timely compliance of procedural requirements of site selection, project preparation and sanction.
- Better extension methodology to increase the pace of crop diversification to more value-added crops with livelihood potential.
- More efficient use of the conserved water through deepening of wells and construction of new wells and also use of drip and sprinkle technology.
- Sustainability of community lift irrigation schemes.
- Convergence of selected programme components/activities with commonality in approach.
- More emphasis on community mobilisation.
- More emphasis on zone-wise preparation of technical recommendations, methodology of extension, and replication of technology and methodologies used in watershed development to get farmers' participation on a larger scale.
- More emphasis on livelihood approach by developing linkages for new non-farm livelihoods—poultry, fisheries, dairy, food processing, rope weaving, horticulture, etc.
- Rationalisation of unit cost norms depending on the nature of programme content, work items and institutional arrangements.
- Feasibility of territorial delineation to be decided in terms of eligibility criteria, ministerial mandate, programme focus and development objectives.

- Scope for enlarging the process of capacity building involving local bodies (PRIs), non-governmental organisations (NGOs), community groups and extension functionaries.
- Broad basing of financial resources through inter-institutional and credit linkages.
- Unified approach supporting programme measures and building suitable institutional framework for ensuring long-term sustainability.
- Enhancing the skills and knowledge of the watershed community by imparting skills and information about value-added activities like nursery raising, horticulture, agro-forestry, floriculture etc.
- Providing opportunities to SHGs of women, rural youth and landless labourers to develop livelihoods based on the livelihood potential created.

Gandhi Gram Yojana

The watershed approach is of greatest relevance to Rajasthan given that 31 per cent of the total geographical area is wastelands. Taking inspiration from the success achieved in Ralegaon Siddhi in Maharashtra, the Government of Rajasthan launched a similar scheme with emphasis on convergence of benefits flowing to the watershed schemes. A unique village-based watershed and community development project has therefore been designed under the title 'Gandhi Gram Yojana', with one village chosen in each of the 237 Panchayat Samitis, based on unopposed election of the Sarpanch, to avoid usual election. The Yojana involves convergence of voluntary labour, family welfare, education of the girlchild, tree plantation and improved animal husbandry practices. The progress under the Yojana up to July 2003 on the five parameters is as follows:

Family planning	
Eligible couples identified	36,755
Couples adopting family planning technique	10,765
Couples adopting contraception	12,990
Voluntary labour	
Total mandays	99,292

Environment improvement	
Forest plantation	3,90,115
Pasture development	1,205
Education of the girlchild	
Identified girlchildren	28,713
Girlchildren enrolled	25,967
Improved animal husbandry	
Campus organised	500
Vaccinations	75,578
Castrations	9,729
Formation of SHGs	
Illiterates identified	37,857
Illiterates linked with literacy programmes	5,362

Recently, some of the social infrastructure required and articulated by the watershed committees in the Gandhi Grams has been sanctioned in the seven DPIP districts. For other districts, the PRIs are to provide this infrastructure by dovetailing with ongoing schemes.

Watershed Plus for DPIP in Rajasthan

Watershed programmes have been implemented in the country under schemes initiated by the Central Government. Many states have taken up development of degraded areas under the watershed approach on a mission mode. NGOs have played a major role in the regeneration of degraded lands in order to increase the carrying capacity of the land and water resources for ensuring sustainable livelihood opportunities and food security for the rural poor. A paradigm shift in the approach to watershed development came in 1995-96 with the adoption of revised guidelines in line with the recommendation of the Technical Committee, headed by Dr. C.H.H. Hanumantha Rao. The new paradigm of 'Watershed Plus' recognises the need to involve the community as a necessary condition for the sustainability of a watershed programme. Watershed development is not just a technical project but also includes a social agenda as well. The inclusion of women and vulnerable groups and a strong focus on equity is what distinguishes the integrated 'Watershed Plus' approach from previous watershed programmes. The programme seeks to ensure convergence of all other programmes that promote economic activities and generate increased employment opportunities. Conscious efforts to promote non-farm employment and increased land access for the landless as well as promotion of self-help groups form an integral part of the new approach.

The revised programme guidelines seeking to operationalise the 'Watershed Plus' philosophy issued by the Department of Land Resources, Ministry of Rural Development, in 2001, provide for a substantially modified approach to make the programme even more result-oriented. The guidelines have recently been further revised and repackaged in the form of guidelines for "Haryali" that gives primacy to implementation by the Panchayats. The key changes that can be adopted in DPIP are summarised as follows:

- Bringing common interest groups (CIGs) of BPL families to the forefront in harvesting the livelihood opportunities created, e.g., nursery raising, horticulture, floriculture, agro-forestry, agro processing, etc., by providing them need-based resources to enhance their livelihood through finances, know-how and technology, access to markets etc.

- Empowerment of the community by providing training of rural youth in skills such as modern techniques of agriculture, horticulture, animal husbandry, construction, etc., so that these skills are locally available and accessible.

- Strengthening the back up through the DRDA/DPMU, WTT block group of experts and village-based para-service 'janakars'.

- Creating backward and forward linkages for selected livelihoods based on local resources and skills to enhance the livelihood opportunities of the community, particularly the BPL. Assistance of NABARD can be utilised for this purpose.

- More emphasis on community mobilisation and improving crop diversification and cultivation of value-added crops in the treated areas, and replication of the technological innovations in the non-treated areas, to move towards the Second Green Revolution.

- More efficient use of water and drought proofing through use of sprinklers, drip and green house technology. More exploitation of the additional water resources harvested by investing in new wells and energisation of existing wells.
- Dealing with environment degradation by stressing on pasture development, ravine reclamation, horticulture, agro-forestry and better animal husbandry practices.
- Food security for the poor by encouraging groups of women and landless to improve their livelihoods and earnings.
- Transfer of technology for farming and also for reduction of drudgery for women.
- Improved incomes for farmers through transfer of technology, ready availability of inputs, and other backward and forward linkages.
- More non-farm employment opportunities for landless labour and marginal farmers.
- Diversification of income generation opportunities for women through CIGs.
- Convergence of services and better access to government schemes.
- Easier access to credit from financial institutions.
- Transparency in implementation.
- Institutionalising and sustaining the watershed committees by registering them as a society.
- Effective use of remote sensing data furnished by the National and State Remote Sensing Agencies for further enhancing the gains from the watershed project.
- Implementation through special Project Facilitation Teams (PFTs) and association of NGOs with implementation, depending on their capacity, and inclination of the CIGs.
- Documentation of innovations in soil and water conservation and their dissemination and practice.
- Additional resources for watershed programmes from all sources.

- More cross-learning from other states, organisations and experience sharing to improve the quality of implementation.
- More effective inter-institutional coordination, e.g., between the government, project implementation agencies (PIAs) and NGOs, and also between the PRIs and line departments.

Key Action Areas

Following are the key action areas:

- Building capacity of the PRIs to implement watershed projects. Training and site visits, posting of support staff, supervision, facilitation in case of delay at any level, etc., are all ways to achieve this goal.
- Holistic approach to watershed development. Build in programme elements of diversification of crops, development of minor irrigation sources, pasture development, tree plantation, develop organisations for women and landless labour.
- Eliminate delays at all levels. Develop a project approach, with automation of paper work, monitoring systems, timely reporting, evaluation work as well as compliance of objections.
- Adequate flow of funds. Tap all resources including foreign donors and elements in other ongoing programmes.
- Harmonise the watershed development with PRIs. Confidence building and intervention wherever friction is reported. Revise job charts to reflect new concerns.
- Field-level monitoring of progress and impact evaluation.
- Improve IEC activities. Compile success stories and develop innovative display materials. Dissemination of information to promote cross-learning in more cost-effective formats is also essential.
- Improve training methodology and utilisation of training resources.

Action Plan for DPIP Plus in Rajasthan

Following is the action plan for 'DPIP Plus' in Rajasthan:

- A DPIP-specific and focused district project approach.

- Placing an advisor, watershed with director, DPIP to assist and strengthen the SPMU.
- Placing an expert with DPMU to monitor and assist in implementation of watershed projects in the district.
- Updating and training of NGOs to take up macro watersheds.
- Setting up a block-level PFT consisting of an agricultural engineer, a veterinarian, a social scientist and an agricultural extension worker where NGO's capacity is questionable or the watershed training team is unable to do the needful.
- Training and empowering village youth to provide services of agriculture/horticulture/agro-forestry; animal health and better husbandry practices; and training retired teachers to work as social animators. Encouraging a group of para-service providers to supplement government services.
- Formulating an MIS to design and collect relevant information as well as devising a reliable concurrent evaluation system.
- Allocation of sufficient funds to implement the project and fund the value-added activities.
- Provide greater flexibility in implementation.
- Well-defined role for state, district and village level institutions.
- Setting up an appraisal mission for the seven districts with four members to specially map achievement differences and remedies and focus on somewhat neglected issues such as empowerment, community mobilisation, training, equity and gender.
- Special emphasis on training and capacity building of the user group and NGO/PFT.
- Programme should be transparent. To supplement the record-keeping and accounting to expenditure on the watershed projects, it is proposed of outsource all difficult skills in a suitable manner to achieve this end. All records should be maintained at the village level. The project should be community-based, rather than secretary-based or leader-based. Regular meetings of the user community must be ensured.
- An exit protocol will be implemented for the project implementing agencies (PIAs). Nevertheless, regular follow-up

visits should be ensured and community-based maintenance stressed by the PIAs and PRIs.

- A twin track approach that provides for short-term and long-term benefits in the implementation of projects. For example, the nurseries could be raised and maintained by beneficiaries on their own fields. The watershed associations could be registered as a society to ensure long-term flow of funds.
- A combination of community organisations/NGOs as PIAs.
- To be more gender-sensitive by ensuring a greater role for women.
- An effective role for the PRIs.
- Building up the watershed in the Gandhi Grams as a need-based holistic and integrated development initiative.
- Nurturing community groups to further long-term benefits from watershed development.
- Establishing credit linkages with financial institutions to ensure sustainability.
- Effective use of remote sensing data furnished by the National Remote Sensing Agency for planning of further interventions.
- Development of audio-visual material and radio programmes for spreading awareness about the utility of the watershed approach.
- Supporting Gandhi Grams in all 42 blocks of the seven districts for overall development as model villages for others to emulate.
- Document indigenous water and soil conservation interventions in Rajasthan and other semi-arid regions of the country to provide a resource base for project implementation in the state.

Care should, however, be taken not to substitute centrally-sponsored schemes (grants) for watershed development with loan funds from DPIP. The approach should be to dovetail both the programmes, utilising funds from the government to develop the watershed as per existing guidelines. Funds under DPIP can be used to augment opportunities for the BPL families so that they can harness benefits from enhanced soil productivity and water conservation. Even potential in completed projects can be utilised to provide additional opportunities for the BPL families and their dependents through interventions and support for SHGs through DPIP.

Expected Outcomes

Following are expected outcomes:

- Additional area under cultivation.
- More area under irrigation—rabi and zaid crop possible.
- Change in cropping pattern—multiple crops and higher value crops possible.
- Increased incomes.
- Increased price of land.
- Higher water table.
- Higher investment in agriculture—seeds, fertilisers and farm implements.
- Dynamism in local economy, and stimulation of innovation and entrepreneurship.
- Improvements in the standard of living.
- Higher yields.
- Contribution to equity in development, large number of SCs, STs, marginal and small farmers benefited.

14

Popular Knowledge, Participatory Planning and Creating a Path to Development: Looking at the Viability of the Methods Used by Tarun Bharat Sangh for Undertaking Water Harvesting Projects in Rajasthan

RAKSHAT HOOJA

Tarun Bharat Sangh (TBS) is a non-governmental organisation working towards community mobilisation for sustainable natural resources management in Rajasthan since 1975.

The initial emphasis of TBS was on the organisation of camps and training programs in order to enrol, motivate and mobilise the youth towards social commitment. Soon, the scope of the activities was expanded to include building of primary education centres (Bal Shalayen) to educate the children of labourers involved in the carpet weaving and mining industries. The canvas of operational also increased and camps were also organised in rural areas.

By 1983, TBS had come in contact with many rural youths and in order to better understand the conditions that these rural people lived in, TBS conducted several studies. One of the most important studies in this regard was a study on the socio-economic development of "Gadulia Lohars" (blacksmiths who moved from village to village in search of work), which brought to the fore the adverse affect of industrialisation on this nomadic group. This and other similar studies highlighted the problems faced by the villagers, the negation of their rights over common property resources, lack of basic facilities like drinking water, lack of irrigation and their inability to fight the system. These studies enhanced the knowledge, skill and confidence of the group in coping with the issues in rural areas and led them to think and evolve a strategy that would be practical and useful at the grassroots level for the common village people.

The findings of these studies greatly influenced the members of TBS, including Rajendra Singh. An employee of the Department of Youth Affairs of Government of India at that time, Rajendra felt strongly about the welfare of village people and the need for sustainable rural development through appropriate utilisation of local natural and social resources. He quit his job, and on 1 October 1985, along with friends Narendra, Satendra, Kedar and Hanuman, took the decision to go to the villages and plunge fully into developmental work.

On the evening of 2 October 1985, the five friends arrived at Kishori village in Thanagazi tehsil of Alwar district. The next day an acquaintance, Sumer Singh of Suratgarh, arranged accommodation for them in Bhikampura village and, using Bhikampura as a base, the group immediately started primary education and health care centres in the nearby Gopalpura village. The interaction with the villagers grew and many residents of Gopalpura regularly visited the little band of workers.

Thanagazi tehsil is located in the semi-arid region of the Aravallis in Rajasthan with the local population consisting mainly of the Meena and the Gujjar communities. It has been reported that 1985-86 were years of particularly severe drought and the decreased water availability and salinity had adversely affected both the problem of drinking water and agriculture. Consequent decrease in agricultural productivity and fodder availability had completely eroded livelihood opportunities leading to migration from the area of up to 70 per cent of the population. The search for solutions to overcome this economic and ecological crisis in the region gave purpose and direction to the efforts of the Tarun Bharat Sangh.

During the drought period the people drew their attention to the breakdown of the traditional water harvesting system in the region. At village meetings, the elders of the village talked about their traditional *johad* that had always been full of water and had ensured a full granary as well. The old *johads* were in a dilapidated state and the villagers did not have the ability and resources to build new ones on their own. Since the people believed that revival of the *johad* would help them overcome the drought, TBS began the work of constructing a johad in Gopalpura village, through voluntary labour. The elders of the village joined in as did people from the neighbouring village, Sillibaodi. Soon, the youth also pitched in, and so the first *johad* was built by the villagers with the assistance of TBS in 1986. Following this, several *johads* were constructed in neighbouring villages, consequently providing employment during the period of drought.

After working in other villages in Thanagazi tehsil like Hamirpura, Bhoato, Koylala, Bachdi, Kishori, Mandalwas, Angari, Samara, Bhuriyavas, Jhiri, TBS expanded its operations to the neighbouring tehsils of Rajgadh and Umrain. By the end of the nineties, Tarun Bharat Sangh had expanded its operations to over 10 districts of Rajasthan. Today TBS is active in over 2,000 villages of Rajasthan, has won many national and international awards for its work, and its approach is being adopted and copied by others.

The present paper, based on data conducted during a two-year long process documentation study of TBS undertaken by the author during 2001-2003, aims to highlight how the TBS model has evolved by using popular knowledge, large-scale public participation and zealous belief in the superiority of a self-created development path.

Using Popular and Traditional Knowledge as the Cornerstone of Development

When TBS started working on water conservation and management issues in Alwar they deliberately chose to use the traditional and popular knowledge available with the villagers. It was because of this reason that they decided to go with an ad hoc localised methodology which the villagers were comfortable with rather than the top down systematic approach that the government was trying. The belief among the TBS workers is that their methodology provides greater confidence and empowerment to the people.

Though the work of Tarun Bharat Sangh has become famous for the revival of johads, there are actually four important structures that Tarun Bharat Sangh revived in the region [which have also been described in detail in Abhishek Sharma (2002)].

The first and the most major structure in terms of water holding capacity are the 'Anicuts'—a stone dam on a river/*nallah*, they are built on common land plugging the main *nallah*.

The second type of structure is the 'Paal'. They are built on private land on the fields of the farmers and their main purpose is to conserve moisture for the rabi season. The need for irrigation in the rabi season reduces drastically with paals. Like the anicuts these structures are also built plugging the main *nallah*.

Table 1

Types of Structures Built by TBS in the Villages Studied

Village	Johad (Old+ New)	Bandh/ Dam (Old+ New)	Ani-cut	Medh-bandi	Farm Pond	Tanka	Well	Pond	Nala Plug	Total
Hamirpur	11 (9+2)	13 (12+1)	3	22	—	—	—	—	—	49
Bhawta-Kolyala	8 (3+5)	4 (3+1)	2	24	—	—	—	—	—	38
Kacherheda	1 (+1)	—	—	—	9	—	2	—	—	12
Khajoora	—	1 (+1)	—	—	13	—	—	—	—	14
Jaisinghpura	—	—	—	—	—	Seve-ral	—	—	—	—
Duharmala	7 (3+4)	—	1	—	—	1 (1+)	—	—	—	9
Kraska	5 (3+2)	—	—	—	—	—	—	—	—	5
Deori	8 (3+5)	4	1	—	—	—	—	—	—	13
Nayadera	—	—	—	—	—	2	(+1)	2	—	4
Suratgarh	9	11	3	18	—	—	—	—	6	47
Neemi	4(2+2)	—	—	—	—	—	—	—	—	4

The third and the most important type of structures is the 'johads'—an earthen concave shaped dam and pond to store rainwater run-off. These are built on common land on the foothills of the hills and are much smaller than anicuts. Unlike the anicuts they are not built on the *nallah* and instead collect water from a small catchment.

The fourth type of structure is 'Medhbandhi' (or field boundary bunding) constructed in the farmer's fields. This again is on private land and involves raising the sides of the fields to store rainwater in them. Such medhbandhi is less useful than contour-bunding from the point of view of water conservation. But contour bunding has not caught on at all in the area.

Other structures made by them includes Tanka (tank). Khet Talai (farm pond) and Bandha (small earthen dam).

Table 2

Number of Structures Constructed in Villages under Study

Village	Tehsil, District	Total Structure	Community	Private
Hamirpur	Thanagazi, Alwar	49	16	33
Bhawta-Kolyala	Thanagazi, Alwar	38	11	27
Kacherheda	Bamanwas, Sawai Madhopur	12	02	10
Khajoora	Sapotra, Karoli	14	01	13
Jaisinghpura	Thanagazi, Alwar			Several field-bunds
Duharmala	Thanagazi, Alwar	09	08	01
Kraska	Thanagazi, Alwar	05	05	00
Deori	Rajgarh (Tehla shetra) Alwar	13	09	04
Nayadera	Nadoti, Karoli	04	03	01
Suratgarh	Thanagazi, Alwar	47	17	30
Neemi	Jamua Ramgarh, Jaipur	04	04	00

Processes and Working Methods of TBS Functioning

TBS initiatives in land-water management and drought proofing, in theory, follow a pattern that is common for all villages. The entire process of TBS intervention as it is supposed to work can be categorised into a few steps. At each step TBS takes up a series of activities.

The main steps of TBS's professed approach are depicted in the flow diagram below:

Diagram 1

Process Flow Diagram of Activities as They Should Occur

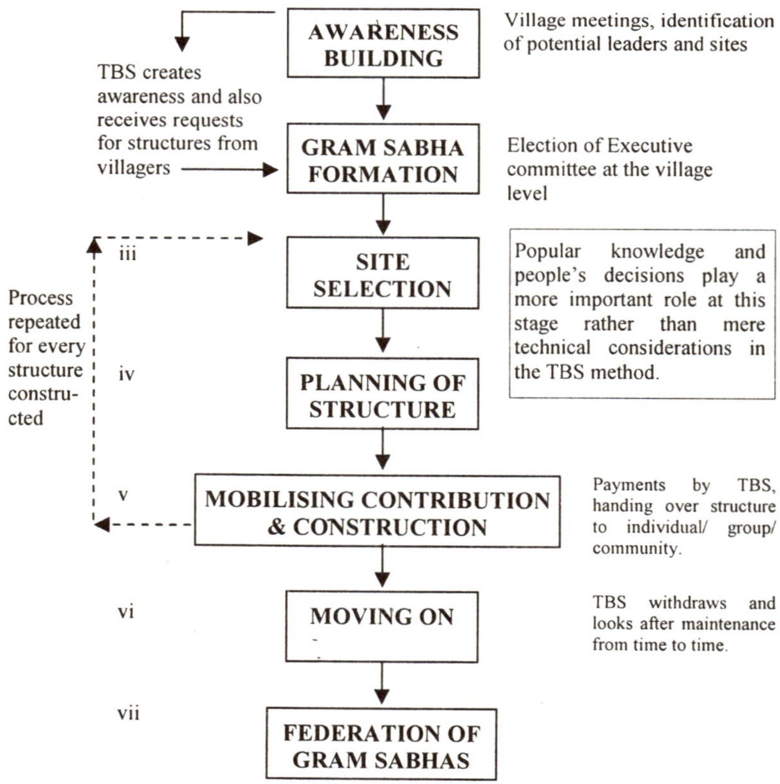

We now discuss each of the steps depicted in the process flow diagram above.

(i) Creating Awareness about TBS and its Mission

The first step taken by TBS in new villages is to undertake an intensive awareness campaign among people about its mission and activities. A common practice followed by TBS is to take the help of locally influential persons who are either known to TBS workers directly, or through persons from the old TBS villages, for their awareness drives. These persons act as contact persons for TBS and help in spreading its message to their relatives and friends. The word of a known person, particularly a relative, has immense credibility in rural Rajasthan. TBS has capitalised on this to establish rapport more quickly and easily. Contact persons thus provide TBS with a credible reference in new villages approached. People who do not otherwise know about the organisation are thus are more easily convinced to participate in its activities.

> **Box 1**
>
> In Hamirpur village, for example, Ruda Ram came to know about TBS work from his relatives in Gopalpura village where TBS was working. He went to Bheekampura and brought Rajendra Singh to his village in 1986. The first work done was on his private Medhbandi and later on other work was taken up in the village. It was through private construction that TBS was able to gain the confidence of the villagers to contribute towards and participate in community work. Since then 49 different works have been done in the village out of which 33 are private and rest are community.

TBS uses different tools to establish contacts in new villages. These are *padyatras, shivirs,* banners and posters, cultural programmes, social networks and word of mouth.

> **Box 2**
>
> A *sant sammelan* was organised in Bheekampura to create awareness among religious people on the importance of water and natural resource conservation. TBS workers felt that the *sadhus* or religious leaders participating in the *sammelan* would be able to spread TBS's message more effectively as they had people's confidence and respect. At the end of the sammelan it was decided that a massive padyatra for water conservation (Akal Mukt Padyatra) would be organised through the state of Rajasthan and it would stop only when the region received its first spell of monsoon rains.

Padyatras (or public march/processions) are the main communication tool used by TBS to make contacts in new villages, build rapport and create awareness regarding the natural resources conservation and water conservation in particular.

Padyatras follow paths that touch villages new to TBS. Villagers participating in the march carry banners, posters and pamphlets with them to spread their water and natural resource conservation message: "*Save water, make johad*".

After the *padayatra* is over, TBS workers hold shivirs to discuss water conservation, drought mitigation as well as the need to do away with social evils such as dowry and gambling-amidst folk songs extolling the virtues of water conservation. *Shivir* are large public gatherings (the word actually means camp) where persons from different villages participate. It is organised in a village where the TBS-village partnership has already performed its activities, where the villagers coming for the *shivir* can also see the results of such activities. They are aimed to build and strengthen relationships with villagers in a social gathering. TBS workers provide a detailed introduction about their activities aimed towards water and soil conservation, drought mitigation and improved natural resources management, and terms and conditions of funding from TBS.

During *padayatras* and *shivirs*. TBS workers look out for promising leaders who will come forward later to take responsibility for forming the Gram Sabha and constructing of water harvesting structures.

Similarly, villagers who are already aware of TBS activities through their relatives or other sources and are interested in constructing *johads* also use such occasions to interact with TBS workers. At times an initial village survey is also done at this time to identify potential sites for johads. In Bhawta, the TBS worker Nanag Ram who had relatives in Bhawta first made contact with the village. Sundra Gujjar of Bhawta village and Dhanna Gujjar of Kolyala were the first people who cooperated with the TBS and took up the leadership of getting work done in their village through TBS assistance.

Box 3

In Kacharheda, Goklendra an influential farmer in the village was the first person who contacted Chaman Singh, the TBS sub-centre head. The first work taken up by TBSsubsequently was to construct Goklendras own Khet Talai in 1997. Later on seeing him many other villagers also got their khet talais made. Goklendra is now the main contact person of TBS in the village.

In other villages like Khajoora, Neemi, Kacherheda, Nayadera, individuals from the village had approached TBS. All those from these villages who had approached TBS were the relatively well-off people of the village.

The awareness regarding TBS activity also spreads through word of mouth, where interested people after learning about TBS activities and programmes approach it to get such work done.

(ii) Gram Sabha Formation

To involve people of the village it is necessary to evolve consensus through participation. TBS came up with its own concept of *Gram Sabha* (not to be confused with the Panchayati Raj Gram Sabha). Gram Sabha formation in villages underlies TBS's concept of a village institution that will ensure participation, involve villagers in the programme, and evolve consensus. The Gram Sabha is meant to decide on water-related issues in the village, to serve as an interface between TBS and the villagers and bring in local knowledge and skills into the construction of *johads*. Before a new structure is constructed, the complete plan, implementation procedure, site, expenditure estimated, amount and mode of contribution and accrual of benefits are meant to be discussed within the village Gram Sabha.

At village meetings, TBS workers introduce the concept of Gram Sabha and how it has to be constituted. People at the meeting are encouraged to identify different persons for the posts of office bearers like the Adhyaksha, Upadhyaksha. Sachiv, Koshadhyaksha (chairperson, vice-chairperson, secretary, treasurer).

The stated TBS policy is to form a Gram Sabha only where other suitable village institutions do not exist. In practice Gram Sabhas were formed even in villages with other institutions. The objective of forming the Gram Sabhas is to make the village self-dependent and to facilitate the decision-making capacities of the villagers. Ideally, as per TBS norms, the Gram Sabha should constitute the following sub committees.

(i) Resource committee for preparing a clay model of the village and for assessing the problems in the village and the views of the community on the size and location of the *johad*.

(ii) Construction committee for overseeing the construction work and the people's participation in the construction.

(iii) Forest committee for enforcing the ban on tree felling and monitoring the regeneration of the forest in the catchment areas.

(iv) Gram Kosh (village fund) to act as a buffer during lean years and to fund the community work in the village.

(v) Grazing committee for prevention of grazing in the protected areas.

(vi) Water committee for management of water resources and sharing of water.

(vii) Women's association for representing the views of women.

In the TBS Gram Sabha ideally one adult male and one female member from each household are to become members. The idea is that each household should be represented. In other cases on member may represent a group of households. During the field study, however, it was observed that the TBS Gram Sabhas are not active in most of the villages and even where they are active all households are rarely represented in the TBS Gram Sabha. It is also not necessary that Gram Sabha formation is done before any physical activities are taken up. In certain villages like Hamirpur and Kacharheda, small private works were started even before the Gram Sabha was formed. At times, it can even take several years to form a Gram Sabha. The form and structure of the Gram Sabha can be quire varied. In Deori, though Gram Sabha was formed in the main village, people in the hamlets meet separately to discuss issues relevant to their own hamlet. The Gram Sabha does not even therefore necessarily represent the entire revenue village. It is also observed that the Gram Sabha at many of the villages have actually acted as construction committee in making of community structure and have dissolved after the construction work has been over.

Box 4

In Hamirpur, Gram Sabha was formed only when a big structure *Jabbar Sagar* had to be constructed in consultation with a large number of people. Prior to this small private and community works were carried out independently.

Even now most work is undertaken without any role of the Gram Sabha. Recently, persons from the 4 hamlets of the village got together sepaerately and raised the contribution for *Dakao Vala Bandit.*

The list of members was available in the minute's registers of Bhawta Koliyala, Hamirpur and Samra. In Bhawta Koliyala the register lists 16 members against 53 households, in Hamirpur 42 persons from 345 households and in Samra 44 persons from 350 households. Except for Bhawta Koliyala, no women were listed as members in the three villages.

(iii) Site Selection

Site selection starts fairly early in the intervention process of TBS. Sometimes, even during the *padayatra* potential sites may be visited. Sites selected for water harvesting are usually of two kinds. Old existing structures needing renovation and sites for construction of new structures.

> **Box 5**
>
> In Ratnoon ka Dera a community *johad* was being made in the forestland. The work started but was later stopped by the forest department and FIR was also lodged against the villagers.

Individuals, groups or the entire community can suggest sites for new structures. Theoretically, site selection is decided on the basis of a few criteria that would help minimise costs. The process may also however be influenced by an influential leader in the village to suit his own requirements. Site selection is ultimately based on agreement between the benefiting community/group/individual/village/dhani (hamlet) and TBS wor-

> **Box 6**
>
> In Deori village, Prabhati Baba contributed to build a *pucca* anicut. According to villagers, his intention was to encroach upon some revenue wasteland, which was at the time occupied by another family. The other family in occupation of the land filed a case against him. Prabhati Baba lost the case and his interest in the bandh. The bandh is now broken due to the lack of maintenance.
>
> In the Jaisinghpura village TBS constructed a private anicut on the government land under the PAWDI project with contribution from Laxman Singh, office bearer in the Gram Sabha. Villagers feel that Laxman Singh actually wanted to encroach upon and cultivate the (sawai chak) government land. The government now wants to make an airstrip in the area under its drought relief programme and so is contemplating removal of unauthorised structure. The anicut itself has become redundant as several field bunds have been constructed in its catchment area and very little water is collecting there.
>
> In Hamirpur, the Arvari Sansad, the village Gram Sabha built from award money received from the President and the Ford Foundation constructed a so called Sansad House. The revenue department has now demolished the building, as it was an unauthorised structure built without prior permission.

kers. In several villages construction on government lands without permission from the concerned department has created antagonism between the Gram Sabha and the government institutions and has resulted in problems at a later date.

Villagers too appear to have many motives—overt and covert—in suggesting sites for public and private structures. In addition to harvesting water, encroachment of common land or prevention of encroachment are common reasons for construction of structures. It may be added that the watershed approach is not always followed by TBS for site selection.

(iv) Planning and Designing Structures

Once site and the type of structure has been finalised, the designing of the structure takes place. The design of the structure depends basically on its type and size and on whether it is *pukka* (concrete) or *kutcha* (i.e., where only earthwork is involved). Villagers themselves design all *kutcha* structures. Where the size of the structure is fairly large or technical support is required TBS provides assistance. Also, design is usually influenced by the direct beneficiaries' of the structure in consultation with local masons and TBS technicians. Along with design, the villagers decide the approximate budget for the structure. For constructing a johad, for example, the budget is based on the amount of earthwork in cubic feet required for the structure. For concrete structures too, an approximate budget is prepared.

(v) Mobilising Contribution and Construction

After the villagers select the site, the households directly benefiting from each *johad* are identified and they accordingly contribute their share of labour or cash.

The percentage and amount of contribution are based upon the estimated budget. The percentage of contribution differs between common structures and private structures. For

> **Box 7**
>
> In Khajoora village a *bandha* has been built with 66% contribution from TBS in comparison to the normal 75% because there is a direct benefit in the form of irrigation accruing to some families and they were willing to put in higher contribution.
>
> The better-off in the village have been able to put in the 50% contribution required for private structures and 13 private *khet talai* have been constructed in the village.

renovation of old structures and other common structures total community contribution varies between 25 to 33 per cent[2] of the total cost. For private structures the beneficiary puts in 50 to 66 per cent of the cost.

The contribution amount may be fixed equally for each household or some may contribute more depending on their economic status.

In Hamirpur, in the making of the Jabar Sagar anicut, which is a new structure, most of the contribution was given from the Hamirpur village and not from the *dhanis* or hamlets of Hamirpur. In the village also the contribution given varied and the Bade Haveli family were the major contributors. Similarly, in making of the Dakao vala bandh, the direct beneficiaries from four *dhanis* came together to make this new structure. No contribution amount was fixed but the beneficiaries raised money in an adhoc and arbitrary manner, 3-4 farmers contributed above Rs. 1,000 while most of the contributions ranged between Rs. 50-500.

For community structures, once rate of contribution is decided villagers hold a meeting and based on the initial estimation of cost start collecting money. Once money is collected, work starts. TBS share of money is a proportion to the amount raised by villagers. For example, if it has been decided that villagers will contribute 25 per cent of cost and they are able to mobilise Rs. 1,000 then TBS will put in only Rs. 3,000 towards the structure. In the case of building a johad where 100 cubic feet of earthwork has been estimated, for example, once the villagers dig 25 cubic feet, TBS will then pay for the remaining 75 per cent. In the Khajoora village, for example, a total amount of 2.9 lakhs was budgeted and villagers were expected to pay over Rs. 70,000. They were however able to raise only Rs. 60,000 in cash. TBS then put in its share of 66 per cent for the structure.

(vi) Private Structures

In the making of the private structure the direct beneficiaries negotiate with the TBS employee for the structures they want and the TBS contribution amount in such structures ranges between 50 to 33 per cent (the amount has been reduced to 33% of late). The kind of private structures are mostly medhbandi, bandh and anicut.

During field research villagers observed that at times contribution has varied from person to person in private structures and persons with more influence have sometimes been able to negotiate a lower percentage contribution.

Villager's contribution is usually spent on buying fuel for tractors/other heavy digging machines—if machines are being used. The usage of machines has to be agreed with TBS workers. Which tractor/machine will be used, for example, will be decided by TBS because skill and experience is required. Fixed cost of renting machinery, cement and masonry costs are usually paid by TBS. Labour and cost of local raw material like stone, sand etc. and their transportation are put in by villagers within their share of the contribution.

Budgeting for structures is based on approximate figures. The total of villager's contribution plus TBS input can at times fall short of the total amount finally required for construction. To complete the structure, villagers then conduct another meeting and mobilise the remaining amount. In some cases, where villagers were unable to mobilise additional funds, TBS did not put in any money in excess of its share—the net result being that the structure constructed was not strong enough. In Khajoora, for example, enough contribution could not be mobilised and as a consequence the Motho wala bandh constructed was said to be weak.

(vii) Moving on or Exit by TBS and Subsequent Maintenance

Once construction of a structure is completed, TBS starts working on other sites in the same village or in other villagers where contribution has been mobilised.

At what point TBS leaves one village to move on is however not clear. In some villages more than 30 structures have been constructed before TBS stops working there and in others only 5 to 8. There is however no comprehensive village level planning of water resources management or integrated treatment of the catchment area.

Within six study villages the main reasons for discontinuation of work by TBS lay in the inability of the village to work together, i.e., on account of conflict or lack of willingness and ability of villagers to put together the required contribution. None of the villages visited have reached a state of saturation, i.e., complete and integrated treatment of the land area.

The field study shows that reasons for TBS exit vary from village to village. At times TBS has withdrawn because of conflict in the Gram Sabha and in others because construction activities have ceased.

It has been observed that in few villages close to TBS headquarters work has continued in phases over many years. In most villages, however, work has been taken up only in one phase after which TBS has moved on. In old villages like Hamirpur, Bhawta Koliyala and Samra, TBS has a constant and continued presence. These villages are very close to Bhikampura, where the TBS headquarters is located. In others such as Nayadera and Kochar ki Dang, work was finished in one phase and even TBS workers have not visited these villages for a long time.

On the question of maintenance, field observations indicate that maintenance of structures constructed with TBS assistance is poor.

By and large villagers have not maintained structures in TBS villages. Only in Nayadera village where tanka water is used for drinking and no other drinking water facility is available, have the villagers themselves taken the effort to clean the tanka on their own without external assistance. In Bhawta Koliyala, the Gram Kosh was used to maintain a few structures.

In a few villages TBS has motivated villagers to maintain structures by putting in part of the money. In Deori and Kraska TBS has put in 66 per cent of the money required for maintenance. The general process for maintenance is that TBS workers, during their visits to villages monitor the structures and take note of those structures that need maintenance. They then persuade the Gram Sabha or group concerned, to raise 33 per cent so that TBS can put 66 per cent. During the study, TBS workers mentioned that where the cost of repairing the structure exceeded Rs 3,000 TBS usually contributes 66 per cent of the cost.

Box 8

In the village of Hamirpur the Dakao vala Bandh got damaged just few months after it was made. The villagers said that this happened because of technical fault as the spill way did not have appropriate depth and due to pressure of the water, part of the bund gave way. And then the villagers together repaired it without the TBS support as immediate attention was required but they are also negotiating with the TBS to reinforce it in the coming season.

The general attitude for maintenance in villages however varies between disinterest and the expectation that maintenance will be done with TBS support. Most Gram Sabhas do not have a common fund that can be used for maintenance and this too is a disadvantage.

Box 9

In Nayadera it was observed one johad has been made in a faulty manner and does not hold water. Water collected in it seeps out from its wall but no steps have been taken to repair the structure. The villagers felt the best possible remedy would be to convert it into a tanka by cementing the whole structure. This they feel will be more beneficial. Cementing is however too expensive to undertake without TBS support. TBS workers have reduced visits to the village and come only once in a year.

Conclusion

The TBS experience in using popular/traditional/people's knowledge and participatory planning for water harvesting shows that though it may not lead to 100 per cent success, interaction of all the stakeholders in both the choosing the tools and the methodology of implementation is essential in any rural development activity. Though maintenance of structures is not up to the mark, water harvesting structures have had an impact on village economics.

After systematically listing out the various stages, components and tools of the TBS methods if we now ask the question whether the TBS approach has been beneficial in terms of change and development in the region, the answer would be a resounding 'Yes' But if we analyse if it has been able to successfully use a combination of popular knowledge and modern technology available to it to create a viable and sustainable solution, the answer is still not clear.

NOTES

1. This refers to person contributing towards the cost of the structure. At times persons not benefiting directly might contribute in the form of donation for a public cause. For common structures the Gram Sabha leaders have taken responsibility to mobilise the contribution and for semi-private and private structures the family head interacts with the mason and TBS person to finalise design.

2. Percentage of contribution for common structures has been officially raised from 25 per cent to 33 per cent during the last two years. For private structures the range is 50 per cent to 66 per cent.

REFERENCES

Agrawal, Anil and Sunita Narain (ed.) 1997. *Dying Wisdom: Rise, Fall, and Potential of India's Traditional Water Harvesting System,* Centre for Science and Environment, New Delhi.

Bouma, Jetske 2002. "Watershed Development in SAT: A Review of Topics for Further Research", paper presented at the Workshop on Policy Futures for Water Resource Management in India: From Prescription and Impact Assessment to Strategic Analysis, organised by IWMI-India Regional Office at ICRISAT, Patancheru.

Farrington, John, Cathryn Turton and A.J. James (eds.) 1999. *Participatory Watershed Development: Challenges for the Twenty-first Century*. Oxford University Press, New Delhi.

Government of Rajasthan 1986, 1994. Statistical Outline (*Sankhyiki Rooprekha*): Alwar.

Government of Rajasthan 1968. *Rajasthan District Gazetteers,* Alwar.

Hanumantha Rao, C.H. 2000. "Watershed Development in India: Present Experience and Emerging Issues", Lovraj Kumar Memorial Lecture, Society for Prevention of Wastelands Development, New Delhi

Hooja, Rakesh 2000. "Administering Water in India in 2050 AD" in *Water and Energy International* Vol. 57, No. 4, October-December 2000, reprinted in Rakesh Hooja, *Multidisciplinary and Participatory Water Management for Irrigation Command Areas.* Working paper/Monograph No. 2, Indian Network on Participatory Irrigation Management (Indian PIM), New Delhi.

Hooja, Rakesh and Ganesh Pangare 2002. "Some Reflections on the Software Aspects of Water Policy: CAD, PIM, Watershed Development and Management and Rainwater Harvesting", paper presented at the Round Table on Water Policy organised by Indian Society of Ecological Economics at IIC, New Delhi, on 10 May 2002, reprinted as Indian PIM working paper no. 13, New Delhi 2003. Also reprinted in Rakesh Hooja, *Management of Water for Agriculture Irrigation, Watersheds and Drainage*, Rawat Publication, 2006.

Hooja, Rakesh and Rakshat Hooja 2005, "The Changing Role of Panchayats in Watershed development" in Y. Sisodia (ed.), *Functioning of Panchayati Raj System*, Rawat, Jaipur, 2005.

Hooja, Rakshat 2002 (a). "The Institution Behind the Magsaysay Award Winner: Tarun Bharat Sangh: A Historical Profile", in *Dialogue Quarterly*

January-June 2002. (published under title "Folk Traditions and Sustainable Development").

Hooja, Rakshat 2002 (b). Revival of Traditional Watershed Practices to Conserve Water through Participation of the Community: Process Documentation and Analysis of the Socio-Economic Impact of the Activities of Tarun Bharat Sangh in Alwar, Rajasthan. Thesis written as part compliance of requirements for M.A. Final Sociology at Jamia Milia Islamia, New Delhi.

Hooja, Rakshat 2004. "Why only Panchayats? Other Local Associations Can Manage Watersheds Effectively" in *Down to Earth*, Vol. 12, No. 18, February 15, 2004.

Hooja, Rakshat and Rakesh Hooja 2005. "Community Based Organisations and Panchayats as Instruments of Governance in the Sphere of Watershed Development", *Prashasnika*, Vol. 31, No. 1, January-June 2005.

Indian Network of Participatory Irrigation Management 2002. Community Water Harvesting: Process Documentation of the Activities and Approach of Tarun Bharat Sangh, an unpublished report by Ganesh Pangare, Talha Jamal and Rakshat Hooja written for Ford Foundation.

Institute of Environmental Studies, Jaipur 1994. Working of Tarun Bharat Sangh, Bhikampura District Alwar Rajasthan: An Evaluation Study.

Indian Water Resources Society (IWRS) 1998. *Theme Paper on Five Decades of Water Resources Development in India*, Water Resources Day 1998.

IWRS 1999, *Theme Paper on Water: Vision 2050*, Water Resources Day 1999.

Jamal, Talha and Ganesh Pangare 2002. "People's Movement for Water: Lessons from the Johad", outline of paper prepared under auspices of Indian PIM for a workshop in February 2002 organised by International Water Management Institute (IWMI) at Anand.

Jamal, Talha, Rakshat Hooja and Ganesh Pangare "Johad versus Watershed: A Case Study of Tarun Bharat Sangh's Approach to Water Harvesting and Conservation for Drought Mitigation" in Jasveen Jairath, Vishva Ballabh and Khalid Hussain (eds.) *Droughts and Integrated Water Resources Management in South Asia*, Sage, India, 2008.

Kaarnik Kusum and Kapoor Anand, *Deori ki Sangharsh Gaatha*, TBS, Alwar.

Kathpalia, G.N. and Rakesh Kapoor. 2002 "Water Policy and Action Plan for India for the 21st Century: An Alternative". Unpublished paper Alternative Futures, New Delhi.

Kerr, John and Ganesh Pangare 2001. "Water Harvesting and Watershed Management" in Ruth Meizen-Dick and Mark Rosegrant (eds.) *Overcoming Water Scarcity and Quality Constraints*. 2020 Vision Focus 9, October 2001, International Food Policy Research Institute, Washington.

Khalakdina, Margaret 1998. *The Promotion of Community Self Reliance: Tarun Bharat Sangh in Action*, Oxfam India Trust, Ahmedabad.

Kumar, Pankaj and B.M. Kandpal 1997. Evaluation Report of Project titled Building 250 Water harvesting Structure (Johads) in Bansur and Umren Blocs of Alwar District of Rajasthan, a report for SIDA, Dehradun, January 1997.

Ministry of Agriculture 2000. *Common Principles for Watershed Development,* Rainfed Farming Division, Department of Agriculture and Cooperation, New Delhi.

Ministry of Rural Development 2001. *Guidelines for Watershed Development (Revised 2001).* Department of Land Resources, New Delhi, November 2001.

Ministry of Water Resources 2002. *National Water Policy*, New Delhi, April 2002.

Oikos and International Institute of Rural Reconstruction 2000. *Social and Institutional Issues in Watershed Management in India.* Oikos New Delhi and IIRR Cavite, Phillipines.

Planning Commission 2000. *Mid term Appraisal of Ninth Five Year (1997-2002).* New Delhi.

Rajender Singh 1995. *Tarun Bharat Sangh: A Decade.* Tarun Bharat Sangh (in Hindi).

Sharma, Abhishek 2002. "Does Water Harvesting Help in Water Scarce Regions? A Case Study of Two Villages in Alwar Rajasthan". IWMI-TATA Water Policy Programme Annual Partners Meet 2002.

Tarun Bharat Sangh Annual Reports (various years after 1995—in Hindi).

UN-Inter Agency Working Group of Water and Environmental Sanitation (1998). *Johad-Watershed in Alwar District Rajasthan.*

Vepa, Swarna Sadasivam 2002. "Importance of Water in Ensuring Food Security: Policy Implications", paper presented at the Workshop on Policy Futures for Water Resource Management in India: From Prescription and Impact Assessment to Strategic Analysis, organised by IWMI—India Regional Office at ICRISAT, Patancheru.

III

Agricultural Development in Arid Areas

15

Some Challenges and Strategies of Agricultural Development and Land Water Management for Rajasthan in the 1980s

BHUPENDRA HOOJA

It is intended in this paper to present a bird's eye-view of the developments to date, and also foresee some of the problems, issues or areas of activity which may emerge in the agrarian structure in Rajasthan in ensuing years.

I

HISTORICAL EVOLUTION AND SOME SHORTCOMINGS

Stray and isolated efforts to create a semblance or infrastructure of a modern society were initiated by some "enlightened" rulers of the princely states, particularly in and around their capital cities, much before Rajasthan appeared on the new political and administrative scene of the country as an integrated unit. However, the real start could be made only after independence and with the merger of nearly two dozen states and chiefships through a process largely accomplished during the hectic days of 1948 and 1949, but completed only by 1 November 1956. We may do well to recapitulate some of the beginnings made in that era.

Representative governments with limited powers and functions were set up in some states on the eve of Indian independence in response to popular pressures and agitations conducted by the local praja mandals in various states. Here and there, schools and medical services had also been started. Some networks of roads and the state railway systems had opened up the interior. In the urban areas, local power houses, street-lighting facilities and water supply schemes had also been started. And during the war years, foundations were laid of some small and medium scale industries.

But, on the rural front, and in the sector of agriculture and allied economic activity, which was the backbone of prosperity and general welfare then, as it remains today, there was no initiative or state effort at all except for relief measures during famines and droughts. The first steps in this direction had to await the installation of a new and democratic or representative political set-up in the new and unified state of Rajasthan.

Among the first, and one of the most popular, measures of the new regime was the series of tenancy and land reforms which broke the back of the feudal and jagirdari system of exploitation, and protected the interests of the tenant cultivators who were brought into direct relationship with the state after many centuries. This gave them some security, status and a stake in the use or improvement of land, which they could treat as theirs for the first time. These measures also fulfilled the demands or aspirations of the rural masses as expressed through various *satyagrahas* and kisan movements since the early twenties.

Simultaneously on the administrative front, various technical and development departments were created or reorganised and expanded with their regional district and field level organisations. The agriculture department was a notable addition among these. To start with, this department looked after multifarious activities or field functions including crop husbandry minor irrigation, soil conservation, animal husbandry and cooperation. It was modelled on the pattern (or guidelines) suggested by the Government of India and its committees or commissions or the set-ups as had then existed in other erstwhile British administered provinces.

As is well known, agriculture and land administrations in the country were not imbued with the ethos of development. These had been evolved during the later period of the colonial rule, chiefly to augment the coffers of the empire, or to link up Indian agriculture to the world-wide but empire-dominated mercantile economy, and incidentally or partly to meet the recurring challenges of droughts and famines.

In the new state of Rajasthan, not only the departmental set-up, but also the package of programmes and schemes or functions evolved by the Government of India or other state governments was adopted *mutatis mutandi* and without much change or adaptation.

Somehow this centrally-sponsored or guided 'universal' programme frame, along with its traditional structure, has continued to prevail until today in spite of the fact that, over the years, due to various types of surveys and research efforts many points of difference or variations from the all-India model or the scheme of priorities in some of the neighbouring states have been slowly identified; and sometimes reluctantly provided for by the policy-makers and development administrators alike.

Even among the farmers in our state, there has been a growing tendency to go in for options or practices or even cropping practices more ideally suited for different and better or more favourable agro-climatic conditions. Until recently, this compulsion or temptation to fall in line with others, or the craze for imitation of a borrowed model or the easy way of adopting ready-made models has worked fairly well. But now it seems that the future agricultural development of the state will have to be more closely related to, and based upon the local problems and potentials and techniques appropriate to their solution, or utilisation of the local and resources.

II

SOME ACHIEVEMENTS AND SHORTFALLS

However, the gains made so far have been quite substantial, even if it has not been possible to maintain the tempo or the rate of growth from year to year or plan to plan. The setbacks have been largely due to adverse climatic conditions, frequent and repeated droughts, erratic or uncertain rainfall and occasionally heavy downpours and sudden floods. Certain institutional constraints have also been there and some have emerged in the course of time.

Yet the overall achievements have been significant, as briefly indicated below:

(i) The area under cultivation has increased from about 93 lakh hectares three decades ago to 145-150 lakh hectares, partly as a result of the satisfaction of land hunger amongst the cultivators who were traditionally deprived of all land rights, and partly as a result of the state government's commitment to allot cultivable government lands and lands declared surplus with the introduction of land reforms and land ceilings to the landless so as to give a fillip to the grow more food programme along with social justice.

However, the expansion in the cultivated area under these measures has worked as a double-edged weapon. Increased pressure on land has disturbed the delicate ecological or agro-climatic balance of natural resources of soils and water. Secondly, though lands have been allotted liberally, and sometimes "unjudiciously" to the landless, without regard to their productive capacity, the beneficiaries have not been helped much with financial or technical support, nor provided with inputs tools and funds needed for proper development of such lands, particularly lands of marginal or sub-marginal productive value. Weaker sections have also felt the pinch of the limited resources available to them.

(ii) The area under irrigation and the harnessed potential of irrigation have also both increased by more than double from 10 lakh ha. to about 30 lakh ha. as a result of huge investments particularly on some medium schemes or major inter-state and multi-purpose projects. Dams and reservoirs and canal systems have been built wherever feasible. The number of wells has also multiplied and with about 2 lakh electrified, and 1 lakh with diesel pumping sets, they are now playing a more than significant part in the protection and increased production of crops.

But the total irrigation potential in the state is limited. Certain other aspects, in fact the real challenges of irrigation on the cultivators' fields or of water management have received comparatively less attention of the authorities, irrigation experts and beneficiaries alike. On the other hand, a style of wasteful irrigation has come into existence, which has created in its wake a host of other problems.

Command area development programmes, reduction or prevention of losses and wastage, enforcement of water discipline and the 'real' pricing of irrigation facilities have been sadly neglected until very recent times.

Even in respect of the traditional system of irrigation through small local tanks or wells, not much has been done to introduce proper techniques and practices for irrigated crops. This lacuna has not been made good even with more and more minor irrigation works or more wells being dug, repaired or improved every year. Efficiency in water-use has not been emphasised, much less practiced, even with nearly 2 lakh wells having been electrified, out of nearly 10-11 lakh total wells in the state, which are supplied electric power at subsidised rates.

(iii) Another general drawback has been that, with millions of cultivators ploughing their scattered and unconsolidated holdings, big or small, the benefits of agricultural research, better inputs or new practices attuned to the local needs or problems have not reached them to the desired extent. The administrative or organisational structure of research-backed extension suitable for such a complex nature of manifold tasks has not yet come into being, in spite of all the talk about community development national extension, panchayati raj and the co-operative way of *sahkari* samaj.

(iv) With nearly 30 and odd years of its existence, the agriculture department of the state or its other complementary and concomitant organisations or agencies cannot claim to have got even a comparable or a reasonable proportion of the number of functionaries which some other departments or government agencies have built up to command and control their limited operations, such as the irrigation department or the department of public works or the large establishment of Rajasthan State Electricity Board or the staff of the medical and health services or the education department and so on. Perhaps even the postal department of the Government of India has many more functionaries in the field than the state agriculture department which has to deal with nearly 45 lakh cultivators and is expected to prescribe or popularise the use of inputs, techniques or practices for as many operational holdings, if not more. Naturally, all extension or demonstration efforts have been very thinly spread out and confined to accessible areas.

There are many noticeable gaps in the spheres of adaptive agricultural research, education and training; and there is a long physical distance or communication gap between the research stations and the farmers' fields which the new or revised extension approach whether under the World Bank-supported project or the ICAR-backed Lab To Land Programme has not been able to cover.

(v) In the arid and sandy desert areas of the state, where conditions are more critical and problems of wind erosion, high and fluctuating temperatures, and droughts are acute, even the limited coverage of research and extension activities provided elsewhere is not available. Serious concern with research effort on dry farming, management of problematic soils and limited resources of water is a little more than a decade old, but its results or findings have yet to be propagated and broadcast. Yet in terms of geographical area, these lands add up to much more than the area of the comparatively more green regions of the state.

(vi) It is also an irony of circumstances that animal husbandry is crucial to the economy of the dry and desert region, and is its main activity (not being subsidiary to agriculture) of the local people. But in the absence of foliage cover and fodder or grass resources, the situation has now reached an almost critical stage. For the state as a whole, in spite of adverse and harsh conditions and in spite of repeated droughts, the cattle population has more than doubled to over 4 crore heads of animals during the last 30 years (see table at the end). All kinds of animal products and by-products like milk, eggs, flesh, meat, wool, skin and hides etc. are being exported outside the state. This may be an attractive proposition in the short-term but animal husbandry on commercial scale is not feasible. In fact, the additional or growing pressure of cattle population may not be in the long-term interest of the state. What is more, the "cultivated area-ploughs-working cattle" ratio is also not favourable in all parts of the state. And there are parts of the state where but for man-power and bullock power, things might grind to a halt.

(vii) Though for the cultivators, agriculture is very much a composite "way of life" with their culture made up of all sorts of rural activities, festivals and rituals linked with agricultural operations, yet at the government or departmental level, each separate activity is the responsibility or concern, as it were, of a distinctly separate department or agency. One agency may look after research, another after extension, yet others after different inputs, or pest control or credit, so much so that an activity like minor irrigation or even soil conservation or development of pastures or land use has been taken out of the jurisdiction and the purview of the agriculture department. This has led to a segregated, rather a fragmented, approach to the rural problems, both at the government level and in the field.

While in the industrial sector, the concept of all services and facilities being provided from a "single window" bas been re-discovered and is being strongly advocated, in the sphere of agriculture the journey has been in the reverse direction. As a "way of life" agriculture did not have many windows, but now various specialities have emerged and experts or specialised agencies have all set up their separate "shops" or compartments, so that the cultivators have no choice but to keep running from pillar to post.

Yet, and in spite of these constraints and drawbacks, the cropping pattern has changed. Crop yields have registered some increase. High-yielding and hybrid seeds as well as more fertilisers are in use; and many tools or techniques of modern agriculture with necessary marketing tie-ups have come into vogue, (as the enclosed table will reveral). The production of food-crops has increased in the last thirty years from about 25 lakh tonnes to an average of between 72-75 lakh tonnes in an average rainfall year. But that is where the hump lies. It has not been possible to meet the decade-old Planning Commission target of 80 lakh tonnes.

These are some other basic or inherent constraints which have now assumed greater significance. These are being discussed at some length in the next part of the paper.

III

LIMITS OF GROWTH

Climate

Except for a few pockets in the south east and south-west corners of the state (in districts like Bundi, Kota, Jhalawar and Banswara, Dungarpur, etc.) Rajasthan lies in the hot, arid and semi-arid zone of the country. In the arid zone, from the western border to the foot of Aravallis, the rainfall ranges between 50 mm to 400 mm per year and is often very erratic. The rainy season is also short in duration. In the semi-arid (or north-eastern) districts, comprising of Alwar, Bharatpur, Jaipur, Sawai Madhopur, Ajmer, Tonk and Bhilwara also, where rain-fall ranges between 400 mm to 700 mm, during the 3 months of the rainy season every year, there is considerable uncertainty about its arrival on time and its continuity through the season, particularly at intervals when various kharif crops need it the most. Since kharif crops cover greater area, the rainfall aspect is very crucial. In this area, as well as in the semi-humid and humid sub-regions comprising of (i) Bundi, Chittorgarh, Udaipur and Sirohi and (ii) Kota, Jhalawar, Dungarpur and Banswara districts respectively, or south and south-eastern regions, kharif is a gamble with the monsoon and rabi crops depend upon the residual moisture of the monsoons or whatever rainfall can be stored in small, medium or large reservoirs.

In the semi-arid and sub-humid zones, due to percolation and somewhat constant and regular recharge of rain water, there are also large stretches of area, some compact and some scattered, irrigated by wells. Out of about 11 lakh hectares area irrigated by wells, nearly 6.0 lakh hectares area is in the seven semi-arid districts, 2.30 lakh hectares in the eleven arid districts, 1.89 lakh hectares in four sub-humid districts and about 0.75 lakh hectares in four humid districts (using the old 26 districts in the state as a basis). Obviously the semi-arid and arid areas depend considerably on protective irrigation and moisture conservation.

Fortunately, the local soils and climatic conditions are also more favourable in some of the areas for a wide-spread use of groundwater through wells for two and even more crops.

Thus, agriculture and allied operations in the state depend upon the monsoon, mostly on summer rains, occasionally supplemented by uncertain and infrequent winter showers. Only 20 per cent of the area can be outside the pale of this cardinal limitation. Other areas can do no better than to have a subsistence level of desert or dry-land based rural economy.

Within the various agro-climatic zones, the basic patterns of cropping and even the established practices of crop rotation have not changed much, except that during the last few decades, with an increased emphasis on irrigation, about 18 to 20 per cent lands have been covered under irrigation. In such areas, there is a greater variety and seasonal variation in the composition of crops because irrigation has been backed by improved technology, credit and inputs for a multi-cropping system (See Table 1 for the area under crops, the irrigated area and the production trends).

By now, however, the state's overall cropping pattern has more or less stabilised itself. Until 1970, foodgrains occupied nearly 78 per cent of the total cropped area. Since then the percentage has been slightly reduced to about 73. Even in the foodgrains group, the low-yielding and "low market-value" kharif crops like bajra, jowar and maize cover large or substantial areas (about 23.4% for bajra. 9.8% for jowar and 4.5% for maize, according to a 1975 NCAER analysis). Dryland kharif crops hold the key to the prosperity of the state and of its rural people.

Irrigation Potential

The irrigation potential in the state is also very limited, even if we include all possible supplies from outside the state through inter-state and multi-purpose projects. Only about one-half of that potential has yet been harnessed.

The Aravallis divide the state into two broad but distinct regions. The major region towards the west and north-west has arid and sandy lands of scarce rainfall and an "enclosed" or "inland" drainage system, except for the Luni river basin. If we exclude this basin and the Gang-Bhakra and RCP Command areas in the north-west and scattered tracts adjoining some local seasonal streams, there is very little surface water potential elsewhere.

However, towards the east and south-east of the Aravallis, there are some fertile plains with fairly satisfactory potential of surface and underground water. These also have agriculturally potential lands where more than one crop can be taken. Districts like Alwar, Bharatpur, Sawai Madhopur, Jaipur, Tonk, Ajmer, Bhilwara, Chittorgarh, Dungarpur, Banswara, Udaipur, Kota, Bundi and Jhalawar lie in this broad zone.

But, just as western Rajasthan can be further sub-divided into three sub-zones according to soil conditions and irrigation facilities, in eastern Rajasthan also, there are three or four distinct sub-zones with the Aravalli range itself constituting yet another sub-zone. Each sub-zone has its own potential of growth, depending upon the rainfall or irrigation facilities and the local soils.

The irrigation potential of the state is estimated to be about 55.50 lakh hectares. But even when this potential has been fully developed in the years to come, the irrigated areas will not be more than 22 per cent or 25 per cent at the most of the total state area. Irrigation benefits are also likely to remain restricted to a few potential areas, like parts of western Rajasthan served by Gang-

Bhakra and RCP systems, the Mahi and Narmada command areas in the south-west and the Chambal area towards south-east, and a few other medium irrigation covered pockets here and there.

This means that any long-term strategy for agriculture cannot depend too much on the doubtful "extra" irrigation potential. However, emphasis needs to be given to all schemes and projects for conservation of moisture, water harvesting and its storage and proper utilisation. Water is one of the most crucial limiting factor for agricultural growth in the state.

Another interesting fact about the irrigation potential is that more than 50 per cent coverage is from the groundwater or "underground" sources through dug-wells and tube-wells, etc, which are generally developed by the cultivators themselves. Most of the wells are in the semi-arid or semi-humid zones and only in small isolated pockets of arid or semi-humid zones where soil conditions are also favourable. Because of uncertain rainfall and recharge possibilities, thousands of wells run dry and have to be discarded every year.

Degraded Lands

Over the centuries, due to wind and water erosion and the long-term effects of over-grazing or continued cultivation and serious but general neglect of the natural land resources and such other factors, the productive capacity of the soils has gradually deteriorated. This is the biggest problem on the agricultural front today.

The problem has also been aggrevated because of the old traditional system of land tenure, which did not provide for a socio-economic climate for much long-range investments on land development, land reclamation or the regeneration of the soil resources. Low soil depths and emergence of salinity and water-logging in different parts of the state and such other adverse factors have also added to the intensity of the problem. This hitherto neglected sphere of agricultural activity now requires early and long-term attention.

For any long-term benefit, the first priority and the foremost task for future development of agriculture will have to be of "land husbandry" or "land management", if we intend to restore or build up or maintain its productivity.

Because of various adverse conditions and other related factors, nearly half of the land area of the state is either not cultivated or is non-cultivatable. On the other hand, during the last three decades or so, along with land reforms and conferment of tenancy rights on the cultivators as also allotment of land to the landless and the small and marginal farmers, the area under crops has gradually increased; but without the necessary public or community-based development efforts and substantial private investments.

At the same time, the uncultivatable half of the state has been completely neglected. This comprises of about 66 lakh hactares of cultivable waste-lands, about 19 lakh hactares of the so-called "pastures and grazing lands" the condition of which belies their name, and about 22 lakh hactares of lands lying unused as "other than current fallow" or unused lands for more than 4 to 5 years as per revenue records. In the years to come, attention needs to be paid to the protection, preservation and development of these lands also.

As per land utilisation data for 1977-78, out of a total geographical area of 3.42,27,185 hectares the net sown area was only 1,51,68,312 hectares and that of current fallows 19,91,134 hectares (combined for two or more crops wherever possible); while barren and unculturable lands comprised 30,06.809 hectares; permanent pastures and grazing lands were 18,25.110 hectares; and land under miscellaneous tree crops and groves not included in net area sown was 9,765 hectares; culturable wastelands 66,04,497 hectares and other than current fallow lands covered 21,92,889 hectares.

IV

INCREASING PRESSURE ON LAND

Over the years, because of a steady (rather phenomenal) increase both in human population as well as in the numbers of livestock, there has been a heavy and growing pressure on the limited land and water resources of the state. It is some consolation though that during recent years, appropriate research measures and management techniques or technologies for improving the productivity of land have also been introduced and adopted at least in some parts of the state by the more enterprising and efficient cultivators.

One result of the increased population has been the increase in fragmentation of holdings and in the number of small, marginal and medium farmers, most of whom have not had (do not have) adequate financial resources, tools or technical skills to properly look after their soil and water resources or to apply in adequate measures the short-term inputs for crop husbandry and long-term measures for improvement. Thus, we seem to have got into a vicious circle which will need to be broken before we can go forward.

We are nowhere near to finding a solution to the inherent problem of maintaining a proper, though delicate, balance between the land and water potential and the needs of the varied plant life, livestock and human population. This can only emerge after a careful and detailed survey of the inventory or potential of the natural resources and then evolving suitable measures for conservation, use and development through appropriate technology.

Fortunately, during recent years, our research and technology have been oriented and developed to a large extent in the right direction and can offer solutions to some of these problems. Steps have however to be taken to translate these research findings or techniques into workable long-term programmes.

V

THE CONSEQUENT CHALLENGES (IN THE AGRICULTURAL SECTOR)

If we identify the different phases of agricultural development in Rajasthan in the twentieth century, the *first phase* would cover the period up to India's independence in 1947 and integration of local states in 1949. This period of stagnation or feudalist status quo hardly witnessed any "development activity" on the agricultural front, though it gave rise to a lot of social discontent and a series of political agitations.

Phase two may be considered to have been initiated in the year 1949, when the first ever steps were taken to modify or change the system of feudal land relations and to provide a series of regulatory, protective and ameliorative measures in the interests of the cultivators in general and tenants or sub-tenants in particular. Simultaneously, steps were initiated to organise the various departments and "service agencies" for agricultural production, veterinary services and animal husbandry, including the upkeep of goats and sheep, poultry and fisheries and also for creating an adequate potential for irrigation in different parts pf the state.

Gradually, in *phase three,* a more clear and defined scientific approach also appeared on the scene. A number of research stations, trial centres, agricultural farms, demonstration plots, or research and breeding farms for improving the breeds of cattle, sheep, goats and camels were established; and new techniques were evolved in the agricultural universities or the Government of India or the Indian Council of Agricultural Research (ICAR) sponsored research stations and also at the state sponsored research units. New seed varieties or fertilisers, tools and implements, new techniques and new practices began to be developed, adapted and spread through community development and NES agencies or through departmental efforts.

Green and White Revolutions

Then came *another phase,* or it may be called a sub-phase, during the late sixties and early seventies, when there was a breakthrough in agriculture with the appearance of at first the improved and then the high-yielding or hybrid varieties of seeds, supported by necessary modern inputs and new techniques with the application of chemical fertilizers and pesticides. This sub-phase or changeover has been generally hailed as the introduction of "a green revolution" (refer to Table 1 for some results of the green revolution).

Subsequently, a similar concerted approach for scientific development of the local cattle wealth was adopted as a part of the so-called "white revolution" which had its slow start in the early fifties and sixties, with the provision of veterinary services, hospitals and dispensaries or the "key village scheme" and other animal husbandry facilities. The emphasis gradually shifted to the introduction of exotic or imported breeds and a market oriented "development" approach with the collection, processing and marketing of milk, meat, fish, eggs, raw and graded wool and such other animal (or waste) products.

The present period may well be the beginning of a yet another challenging phase in the agricultural and allied sector of development activities. The present day status of land use and production data have been indicated in Table 4.

In the *new phase,* and perhaps up to the year 2000, the biggest challenge that the people and the state administrators, scientists and policy-makers may have to face, alongwith the rest of the country, would be to ensure that our state or national food basket (containing all the various items of essential and supplementary diet from the point of view of proper nutrition and health of the people) is kept full.

Land and Water Management

Our population has been growing at the rate of 26.20 per cent, 27.81 per cent and 32.97 per cent respectively during the last three decades, and is likely to grow at a fairly steady rate. But, unfortunately, it is no longer possible to bring new areas of cultivable land under the plough or to go on increasing the areas under assured irrigation or even to find new lands for pastureage and grazing or lands for growing shrubs or trees of for fodder or social forestry etc. As it is, already the precarious balance of land use for agriculture has reached a point of near-crisis beyond which it cannot be expanded or increased indiscriminately. On the other hand, lands used for other purposes or those not under the plough are also not being looked after at all.

The challenge is how best to utilise the existing land resources, and I may add to this, the problem of utilisation of the available water potential, so that the necessary plant life, livestock and human population may continue to draw the necessary sustenance without damaging or deleting the available resources.

Ecological balance must be restored and maintained. We cannot draw more from the natural or physical system than it can supply. Instead of physical or linear growth or expansion in terms of geographical area, attention has to be given towards the necessary qualitative changes in the disciplines of land use and application of appropriate techniques for soil conservation, land development or reclamation, water storage, development and utilisation and the like.

The approach or the objective should be, "from each unit of land or water according to its capacity and to each unit of land water, inputs or preservation technique, according to its needs." That is the secret or the long range development of balanced and sustained development. The emphasis has to be on optimising the output per unit of land or per unit of water but within the possible range. This is a challenge not only for the research scientists or extension staff, but also for policy-makers and the people at large.

Brown/Yellow Revolution through Dry-Farming

Since the scope of increasing the land area under cultivation or the area under irrigated crops like rice, sugarcane or cotton during *kharif* and wheat, barley or gram etc. during *rabi* is very much limited, we have to think more and more in terms of dry-farming techniques and strategies involving a suitable package of practices for traditional and hardy crops like *bajra,* millets, pulses, gram and oilseeds etc. which can be grown in the hot and dry agro-climatic and soil conditions of Rajasthan.

Rajasthan produces about 10 per cent of *bajra* in the country, 11 per cent of oilseeds, 22 per cent of grams and 30 per cent of barley (Table 2). These potentials and well-established "plus points" need to be fully exploited. In other words, what is now needed in a state like Rajasthan is a sort of coarse grains pulses and oil-seeds based "Yellow or Brown Revolution" fully backed with research, extension. inputs support and incentives to cover the development of coarse-grains and all these crops which are "yellow" or "brown" in colour and which require comparatively much less water. These crops which cover the greater part of the cultivated area alone can bring the sunshine of prosperity to the arid or semi-arid regions and assured incomes to the weaker sections, small, medium and marginal farmers or those who are not blessed with the boons of rich soils or benefits of irrigation. Eventually, the "Brown or Yellow Revolution" can also contribute to the objectives of social or distributive justice.

This aspect or potential of development has been neglected over the last three decades or so, except that, only very recently the Government of India, the Indian Council of Agricultural Research (ICAR) and the state government have become alive to the needs as well as benefits of developing these dryland crops, including coarse-grains, pulses and oil-seeds. We have yet a long way to go in the spheres of research, extension and spread of knowledge and mobilisation of inputs for promotion of these crops and achieving optimum results.

Water Discipline

In respect of the irrigated potential which has been created at such a huge public cost, the most important aspect or challenge is to ensure that there should be optimum utilisation of this potential through "proper discipline in water management" by the beneficiaries themselves, and through "command area or on-farm development works".

The area under irrigated crops can be extended all over the state, particularly in the commands of major, medium and minor projects and under irrigation by wells, if water is applied strictly as per crop requirements, and various water saving, storage and recycling measures are adopted through water-course lining, pondage and drainage of fields and the command areas where necessary, and micro or local level lift irrigation, sprinkling and drip schemes.

Cropping patterns need to be reviewed in favour of crops needing less water so that the frequency or intensity of water application may be reduced, and irrigation facilities extended to adjoining new areas or to cover equally worth-while crops needing less water elsewhere. During the *kharif* season, maximum use of irrigation potential should also be ensured, where such potential has been created and soil conditions are favourable.

Animal Husbandry and Fodder Development

In Rajasthan, animal husbandry is an essential and integral part of agriculture with about 40 lakh bullocks, working buffaloes and camels being extensively used for ploughing, for drawing water from wells, for irrigation or for human consumption and for various other operations like threshing and carrying produce to the market. Besides, cows, buffaloes and goats provide milk as an important ingredient of human diet; and other animals like sheep and goats add to the rural incomes with their numerous products and by-products. It is but natural, and all the more necessary, that this sub-sector must continue to receive due attention of all concerned.

Fortunately, over the past few centuries when gradually our economy must have changed from the pastoral to a predominantly agrarian character, many local but hardy and useful breeds of cattle, sheep, goats and camels have emerged and proved their worth in the harsh and inhospitable environments. These should be preserved and improved through selective breeding, proper health care and proper feeding. The recent tendency to replace them with exotic and mixed breeds is of a doubtful lasting value.

Veterinary services and animal health cover schemes have also been developed during recent years almost from scratch. Key village and cattle development schemes and improvement of breeds including cross-breeding, have also played their part. And more recently, attention has been focussed on improved feeds and fodders, and the commercial potential as well as management aspects of the livestock. The state is endowed with some nutritious and good quality grasses, bushes and trees with nutritions folliage, though these are not cultivated. But the rate at which some species of animals have been growing in numbers inspite of their seasonal export and sudden loss during natural calamities including droughts, has now begun to pose a serious problem about the pressure of animals on the available land and water resources, particularly since large-scale migration of animals during famine years is no longer possible; and within the state, many erstwhile open or vacant lands have been brought under the plough.

It is, therefore, necessary that the development of livestock and the raising of fodder crops or useful and nutritious grasses, fodder bushes and trees with lots of leaves should be incorporated in any future programme for agriculture or in any comprehensive policy for land and water use. It is also necessary that all suitable lands which have escaped the plough (or urban and industrial development) should be taken-up for raising fodders, grasses, bushes and trees. In fact waste lands, lands other than current fallow and the so-called pastures or grazing lands, and even the reserve forests, should be brought under such a comprehensive programme of management so as to add to the fodder resources of the state. At present, many such "problem" or "potential" land areas are outside the scope of private, group or community and state management.

For example, there are about 30 lakh hectares of barren and unculturable lands, about 1 lakh hectare each in 14 districts and nearly 4.35 lath hectares in the hilly Udaipur district. A lot of natural vegetation can be preserved on these lands just by fencing and such other protective steps. Out of 66 lakh hectares culturable waste-lands, nearly 50 per cent are in Jaisalmer and 25 per cent in Bikaner district. Five other districts have 2.0 lakh hectares each and one district 1.0 lakh hectare. Some suitable use with an appropriate strategy of development can be found for such lands. May be we can develop them as cattle sanctuaries or fodder reserves for the drought years. The same about "other than current fallow lands". If need be, such lands can be earmarked or reserved for livestock or developed otherwise alongwith pastures and grazing lands which cover about 15 lakh hectares today but are being neglected by the panchayats to which they have been entrusted. Panchayats and the state agencies concerned cannot afford to be lax in the enforcement of discipline.

During the last thirty years or so, the livestock population in the state has nearly doubled; and it is now more than 4.0 crore, the growth during 1972 and 1977 Census being about 6.5 per cent. Meanwhile, as briefly mentioned earlier, more and more lands, including marginal or sub-marginal lands, usually left open for the animals to roam and graze about in the past have also come under cultivation. In the command areas of irrigation projects, even old pastures have also been similarly cultivated. On top of it, the monsoons have also not been quite favourable.

No wonder, therefore, that according to some estimates even in a good rainfall year, fodder of all types is sufficient only for 80 per cent of animals in the state; and during lean and bad years, only for about 45 per cent. We must either augment the available fodder base or drastically reduce the number of livestock, at least by getting rid of the useless and burden-some animals. There is no other choice.

This search for implementation of a suitable (or practical) programme deserves an overriding priority, even before we venture further and enmesh ourselves into more ambitious commercial and market-oriented programmes relating to animal husbandry such as milk production and collection or marketing of dairy produce, export of live animals on the hooves or meat, or hides and skins, and the grading or sale of wool.

There is an inherent and inbuilt relationship between the proper or scientific and judicious management of land, agriculture and animal husbandry. At least, it was so in the old pastoral or even feudal agricultural systems. That is what had sustained our more or less self-sufficient, self-contained and self-managed village communities through the centuries of political change. We have now, once again, to restore that interrelationship.

Perhaps the key to such an effective arrangement lies in identifying the "land use capacity" of any physical area in its environmental system and treating each such piece of land, tract or zone to the measures of improvement or burdens it can cope with, and no more.

A series of technological measures or combination of techniques relating to various specialities of agriculture, animal husbandry, soil and water management will then have to be thought out, recommended and applied as per needs or potentials and problems of the local areas. But, it will not do to pose animal husbandry as an *alternative* to agriculture or vice-versa. as is sometimes done by advocates or enthusiasts of one lobby or the other.

VI

TOWARDS MODERNISATION

The overall and main challenge, particularly in our rural setting, is that of modernisation, viz., application of new technology and building up new institutions in keeping with the best of our old traditions and the new social and economic objectives.

(i) The natural "limits of growth" also impose some limits on the "import or application of technology". For, example, mechanisation, tractorisation, electrification, even irrigation and the application of various sophisticated inputs and techniques have their limited role or scope in the state.

Many of our soils and social institutions (and conditions) will not respond favourably to these measures, unless they have been adapted, tried out and demonstrated successfully before being recommended for wider use. Besides the local bullock power, cow dung compost and farm manure cannot be easily replaced. Thus, the manner and pace of modernisation cannot be the same as elsewhere, say in the Punjab or Haryana.

Rajasthan has to evolve its own model, its own appropriate technology and its own strategy of development which, as we have seen, must need be comprehensive or all-embracing as far as activities connected with the use or development of land and water resources go.

(ii) Secondly, it has to be ensured that suitable practices or techniques and inputs reach all sections of our rural society, in particular the actual cultivators and cattle-breeders. These practices cannot and should not be evolved or advocated only for the affluent and well-to-do or the more vocal, organised and powerful sections, as has been happening hitherto. For this, it may now be necessary to identify and recommend various permutation and combinations of techniques, tools, inputs and land management measures or agricultural practices not only to suit different climates or soils but also to suit different sections of cultivators and their pockets.

Some such efforts have been made, though not very effectively or successfully, under the recent (1974-82) World Bank supported project for agricultural research and extension with the experts, specialists and extension workers recommending suitable farming practices according to local soil and climatic conditions but also according to the means or the inputs available with the "contact farmers" in 12 districts, and not insisting upon the use of the more sophisticated, costly and often not easily available inputs. This approach may be extended to the dry-farming areas also.

(iii) For ensuring that the benefits of science and technology and the various incentives sponsored by the government reach all sections of the rural people, it is necessary not only to expand the existing research, demonstration and extension networks, not only to look into and reorganise the marketing and distribution set-ups, not only to extend the credit and services sectors, if possible through co-operative channels, but also to see that the beneficiaries (cultivators and cattle-breeders) become real participants in these measures and are thus actively involved in the programmes, in particular the small and marginal farmers or animal breeders. The last man must come first. His small plot of land must receive due attention of the extension worker and the manager or supplier of inputs or subsiders.

(iv) It is also imperative, as has been perhaps the main theme of this paper, that all rural development programmes must be "land based" or co-related to and integrated with the programmes for preservation and development or management of land and water resources. Such has not been the case even with some new schemes of individual benefit or assistance like SFDA, MFAL, IRD and the like, though the New 20-points Programme has once again focussed some attention on the need for improvement of land.

For Rajasthan, this focus on land resources must precede all other activities. In fact, "comprehensive area development programmes" need to be taken up all over the state covering both the areas of agricultural potential as well as areas more suitable for livestock and dry-farming. Benefits for individuals and groups like the weaker sections scheduled castes or scheduled tribes should be linked with strategies of area development for more effective and lasting results.

(v) These programmes can be accelerated if some reforms are also brought about in the agrarian structure. For example, consolidation of holdings, which has been long postponed for fear of alienating some sections of the cultivators, is long overdue. Years ago, in the Punjab, this programme proved a harbinger of change and progress. Similarly, it is necessary to eliminate or at least reduce the hold of absentee landholders or those who do not actually live on their lands and do not personally cultivate (or supervise) the same. It is the considered view of this author that such a measure would be more effective for land improvement than the anticipated benefits from land ceilings, at least in the setting of Rajasthan.

(vi) Because of such social constraints, the actual tillers cannot take the risk of substantial investments on land improvement. Easy credit for such capital works is also not available. What the co-operative land banks manage to provide is but a small fraction of what is needed. Perhaps the best thing will be to evolve suitable "watershed projects for comprehensive land use and development" duly backed by technical guidance and credit assistance.

(vii) Apart from good rainfall and irrigated lands, Rajasthan is a single cropped state. In view of the variations and vagaries of monsoon, it is also necessary to involve some variable or flexible cropping pattern or packages of practices and inputs, the package for normal monsoon being different from those required for early or late monsoons, or when rains come with long gaps in between. The Central Arid Zone Research Institute, Jodhpur, has made some studies and recommendations in this regard.

(viii) And perhaps the time has come when some sort of "crop insurance" is introduced in Rajasthan to cover risks of crop failures or damages due to natural calamities. Crop insurance premium can be collected with the land revenue. This may bring about a measure of sustained effort and stability in the regime of rainfed dryland crops. In the absence of such a protective cover, dry land farmers do not use all the inputs.

VII

A BROAD SUMMING UP OF LAND USE DATA

During the last three decades of the development era, cultivation has been extended to wider areas so that many marginally or sub-marginally productive lands have been brought under plough. According to the 1976-77 Agricultural Census Data nearly 57 per cent of the geographical area of the state is now under use (this includes current fallow lands also). Most of these are single-crop lands and depend on the timely arrival and adequate coverage of monsoon rains.

Land Use Data as per 1976-77 Agricultural Census as Compared to 1970-71

Land Utilisation	1970-71		1976-77	
	Area '000 hectares	Percentage	Area '000 hectares	Percentage
1. Net area sown	15395	75.7	14085	70.7
2. Current fallow	1567	7.7	2212	11.1
Net cultivated area (1+2)	16962	83.4	16297	81.8
3. Other uncultivated land excluding fallow land	146	0.7	258	1.3
4. Fallow land other than current fallow	1464	7.2	1412	7.1
5. Culturable waste land	1559	7.7	1742	8.7
Total uncultivated land (3+4+5)	3169	15.6	3412	17.1
6. Not available for cultivation	210	1.0	218	1.1
Total	20,341	100.00	19.927	100.00

N.B. These figures relate to land area occupied or under operational holdings. Fluctuations indicate rainfall and moisture variations.

But for the uncertainty or failure of rains, perhaps more lands which are now being left fallow for nearly four to five years may also have been brought under plough. Not only the area of cultivation has been extended, but increasingly more intensive effort is being put through irrigation, mechanisation, investment on land improvement and costly inputs like seeds, fertilisers and pesticides. But there is always the danger of the soil and moisture resources and the available nutrients of the land being over-exploited, accompanied by hazards of erosion.

VIII

MANAGEMENT OF COMMON LANDS

The data for land use do not reflect the nature or quality of use. While in respect of cultivated lands we may be able to draw some satisfaction from the assumption that all land-holders will look after them out of personal self-interest, about other lands and particularly common lands or waste lands we cannot have that satisfaction. Neither there is a government agency nor any responsible body of local villagers fully concerned with the upkeep and maintenance of such lands which are going from bad to worse.

To my mind, the only satisfactory and responsible organisation which may eventually emerge and become responsive to the needs or problems of the local areas can only be an appropriate form of a "people's organisation" or "a beneficiaries group". We may call it the panchayat or a mandal panchayat or the panchayat samiti. But whatever its area or unit, it should have all the necessary technical expertise and financial or managerial resources at its disposal. It must enjoy the confidence and trust of all sections of society. All local lands and local water resources should be left to its management. Only in the case of gross mis-management or negligence, the custody of such common lands may be taken away from the people's organisation (panchayat, panchayat mandal or the panchayat samiti as the case may be) by a government agency or manager. However, a long-term programme of land use must be prepared and strictly adhered to.

In fact, I would go a step forward further. Not only the so-called "common" or uncultivated lands, waste-lands, ravines and river beds, but' also the local grazing and pasture lands (and later on forests, as well as the; cultivable lands, particularly other than current fallow lands) can also be placed. under the care and custody of the local people or their panchayat, mandal panchayat or the panchayat samiti, depending upon the area of a natural physio-geographical or topographical unit of which such lands may constitute a natural watershed.

This brings us back to the modern concept of land management.

IX

WATERSHED MANAGEMENT: FIRST STEP TOWARDS REGIONAL DEVELOPMENT

It has now come to be accepted that watershed management is the proper approach towards the conservation and building up as well as scientific and continued use of the potentials of our land and water resources. It is also being increasingly realised that, for any organisational arrangement for proper watershed management, the active association of the beneficiary participants in an institutionalised and organised manner with sufficient financial, technical and managerial resources at their disposal is very necessary.

As suggested earlier, the management functions should be entrusted to the participant beneficiaries with the assured support and participation of the technical, professional or financial agencies of the state or other institutions. Unless the people are made responsible, they will not become responsive and vice-versa.

What is required, first of all, is the basic survey of the land and water resources and a technical appraisal of their ultimate potential. Then comes the application of the proper technology, which should be viable and economical also, through certain inter-connected and mutually supporting services. Such a programme of development has to be backed by consistent and long-term public policy to be carried out through the people's representatives with the full support of the administrative and technical apparatus of the state.

Recommendations of a National Seminar

At a National Seminar on Watershed Management held at New Delhi in January 1980, the demarcation of regions, basins, catchments, sub-catchments and watersheds as suggested by the All-India Soil and Land Survey Organisation was accepted, and for the watershed, a notional area between 1000 to 5000 hectares was proposed, with the recommendation that the area should be flexible according to local conditions (in plains and almost level ground areas, the watershed may be of a larger size). Below the five-stage national demarcation, four further small sizes were also suggested to be called "sub watershed", "milli-watershed", "mini-watershed" and "micro-watershed" respectively.

It was also recommended that, at the time of planning and implementing the programmes of watershed management, the natural boundaries of watersheds should be followed for items relating to the development, conservation, storage and use of water; and for other activities like agriculture etc., the existing "revenue village boundaries" may be used.

The smallest viable units, that is the micro watersheds, may be taken up for development as separate units or in the form of manageable clusters or such units, where such units occur in close proximity to each other.

Whether the programmes, which should be comprehensive and complimentary, are taken up at the "micro-level" or at a small or big watershed level, we cannot lose sight of the fact that the smallest unit as well as the intermediate or bigger level should really be considered as parts of an "integral whole" which may be a "catchment area" or a "river basin" or an "agro-climatic or ecological region". This suggests the adoption of a long-term strategy of area plans, or regional development programmes for development and conservation of natural resources.

The approach will have to be more comprehensive and multi-disciplinary, the experts and functionaries of various disciplines or services being placed under a common organisational setting at each different level of planning and programme implementation.

However, what is being proposed is not something entirely new because a similar approach was the basis of the original community development and NES set-up of rural development. The same approach has again been repeated with slight modifications in the programmes of research and national extension which have been launched with financial support from the World Bank since 1974. The CAD and Special Areas Programmes are also being administrated along these lines.

Some Conditions for Success

Land management for cultivation or any other purpose on a watershed basis will be successful if some, not all, of the following conditions are met: (i) If care is taken that the size or area or of the land unit is large enough to generate the required resources and in which the different types of land use can be planned on an inter-dependent and multi-disciplinary basis. (ii) The area should be viable, economical and suitable for the application of various technologies. (iii) At the same time, the community of the land-users or beneficiaries, whether cultivators or cattle-breeders or both, or agricultural labourers should also be as closely knit as possible. (iv) The community should be properly motivated and organised. (v) There should be easy access to outside help, guidance, credit and other inputs. (vi) If management leads to some marketable surplus of the agricultural or natural produce, by-products or waste products, proper tie-up with the local markets will be necessary. (We can add to this list as we go further into operational details.)

X

AUTONOMOUS UNITS FOR WATERSHEDS

When we take into view these and other related considerations there is no escape from the fact that perhaps the smallest viable, economical and technically efficient unit can only be at the level of the panchayat or mandal panchayat, depending upon the size and area of the local watershed.

For larger watersheds or groups of small catchments and sub-catchments or small groups of what are called "mini" or "micro" watersheds, bigger management units or mini units can be organised, as the case may be.

But the organisational or the administrative set-up will also have to be such that it can co-ordinate and supervise all the multi-farious activities relating to agriculture animal husbandry, soil conservation. water-harvesting, inputs, credit facilities, marketing and so on.

The ultimate and perhaps an ideal arrangement would be to have representative and almost autonomous bodies or institutions of the local people with necessary government support and participation at different local or regional levels.

At present, we are having a slightly different type of field set-up, particularly at the district level, where besides field units of government departments and the cooperative network, organisations and development agencies like SFDA and MFAL and now IDA etc. have been set up, largely through government initiative and under government control.

In the long run, the role of the beneficiaries or the people's representatives and the government agencies will have to be completely reversed.

While the guidance, planning initiative and management etc. should come from the people's representatives, it should be the government's function to provide appropriate and economical technologies or technical guidance and wherewithal, with adequate financial support and assistance in subsequent activities like marketing opportunities, along with transferred or delegated power and authority to enforce necessary rules and regulations.

XI

IRDP STRATEGY AND SET-UP

Recently, a new programme strategy has been initiated under the Integrated Rural Development Programme. Increasing production and productivity in agricultural and allied spheres is one of the key tasks of this programme with which it is proposed to link the centrally sponsored scheme of National Rural Employment Programme, so that the rural poor may have additional opportunities of employment during the lean seasons through work-oriented local projects.

IRDP also aims at development of resources and, in case of the vulnerable or weaker sections, promotion and improvement of their skills, provision of credit and marketing facilities and essential minimum needs of the community and the like. In fact, it is a combination or integration of the essential features of a variety of programmes launched in the past and some of which are on-going. Unfortunately, at the levels of implementation, care and development of local natural resources according to their potential are not receiving the attention they deserve.

Steps have also been initiated to establish suitable organisational machinery, but mostly at the district level, to prepare five years' development projects for selected blocks in a phased manner.

The first starting point of these projects has to be the development of agriculture and allied activities in the primary section, and then take up the potential of development in the secondary or terrtiary sectors and to create or bring together the various services, extension, credit, inputs and marketing agencies.

A single executive agency is proposed to be set up at the district level. At present it is likely to be an official agency. But at the same time, it is proposed to resurrect the old CD or NES block type of organisation and strengthen it with staff for specialised and village level functions. The Government of India has already started giving the necessary subsidy for the IRDP set-up at the district and block levels.

Need for Change

In view of what has been stressed in the earlier parts of this paper, it is strongly advocated that, as soon as possible, the character of the district IRDP agency should be changed to make it more representative and responsive (or responsible) to the needs and urges of the local communities. The district-level specialists of various disciplines should act as their advisors and agents. Ultimately the "district agency" should be able to hire and fire its own team of experts and technicians, rather than remain dependent on the technical and development departments or services of the state government.

The same approach should be extended to the block organisation and the next level below, if any.

Local administration on the concepts and principles of autonomy is the inevitable direction in which we should proceed if all the objectives and targets have to be achieved.

Secondly, with regard to the (five years or longer period) perspective plan of local area development, which is now being attempted at the block level, it is strongly urged that the basic concept or approach of "comprehensive land and water management on the basis of natural water sheds" should be adopted, and if need be, the existing block or even district boundaries may be re-arranged accordingly to accommodate and adjust the natural watersheds or groups of watersheds.

All the necessary steps which have been enunciated earlier, such as survey and inventory of the potential present-day use and future balanced plan with regard to agricultural, forest, grazing and waste lands etc. and the identification of special problems of the area must be included in the preparation (as well as the format) of the local block plans.

Need to Integrate

Lastly, there is need to integrate the various organisational and staffing or servicing models which have been recommended in the context of various development strategies or as per the segmented approach of developing one or the other sphere of activity.

We have the working model of development and technical departments extending their hierarchies to the field level. We have the models being followed under the World Bank supported National Extension Projects, or the dairy projects. We have the models (rather shadow systems) of SFDA, MFAL, DPAP and such other "target population" or target area programmes, which are largely based on the core of district administration. We have the models recommended by the Ashok Mehta Committee on Panchayat Raj and various other state or regional committees. Recently, we have worked out the arrangements under the Command Area and Tribal Area or Desert Development Projects or programmes. Various seminars on watersheds management and the problem of arid and semi-arid zones have also made some recommendation.

Perhaps some of the common features of these models or recommendations are: (a) decentralisation, (b) regional area approach, (c) support of specialist services, (d) people's participation or association of participants with the programmes, (e) state level support with funds, technical guidance, (f) credit facilities, and so on. These aspects need to be provided for in the future or proposed set-ups for land and water management.

XII

FOR LAND MANAGEMENT: POWER AND TECHNOLOGY TO THE PEOPLE

We have to give primacy to the roles of local administration and the local popular representatives. The approach may be summed up into three working slogans or catchy phrases: (i) Right and power to the people (participants) to manage their own local land and water resources, (ii) Science and technology at the disposal of the local people and (iii) Manpower or technical resources besides adequate funds at the disposal of the local people to enable them to manage their own affairs, plans and physical resources.

Over the years, as various technical services and disciplines connected with different aspects or segmented sub-sectors of our rural economy have developed and gained their own momentum with earmarked budgets, service cadres and departmental hierarchies, there has developed an unfortunate tendency to keep each unit of activity and department as separate and distinct from the other as possible. Each field unit has tried to preserve and proclaim its own identity or personality. This has made field level coordination a difficult and almost impossible task. Thus, for example; the distance between the soil conservation organisation and its field units with similar or other corresponding organisations and field units connected with the supply of credit or agricultural inputs or agriculture extension or minor irrigation has been gradually widening until no focal point or common meeting ground for common objectives has been left. This tendency has become all the more prominent in respect of some other sub-sectors such as minor irrigation, animal husbandry, cooperatives and the like, which have come under the protective umbrella of special projects.

XIII

REVISED ORGANISATIONAL STRUCTURE: A PROPOSAL

A possible revised organisational structure for the variety of programmes which have to be taken up now on a more integrated and comprehensive scale, and as part of inter-disciplinary efforts and mutually supporting activities with the beneficiaries playing a predominant role, may be somewhat on the following lines:

(1) At the state headquarters level, for the last two or three decades, the office of development commissioner has been there, though over the years, his functions have been changed and some of his responsibilities transferred to other secretariat level. But he has to be made to function as the chief coordinator or the chief leader of all the organisations and agencies connected with agricultural and allied sectors and rural development, including land and water use.

(2) Either under his chairmanship or separately, there should be a coordinating team of heads of technical or development departments, who should function more as technical experts or staff officers, rather than as managers or controllers of the budgets, services, institutions or staff hierarchies under their

control. Apart from their local area functions in association with the local authority, they will remain responsible to the state government.

(3) Similarly, at the regional district, sub-district and tehsil or panchayat level the role of the specialists and technical functionaries will have to undergo a big change in respect of the programmes and functions and services which would now go under the full control and jurisdiction of the local authorities.

(4) Experts and technical functionaries at the local level may have to (or should) break their links from their present-day parent departments but may be selected or allotted to work as members of a common team at various levels of functioning viz., at the state or secretariat level, and then at Head of Department level, then at regional and district or sub-district level or at the tehsil or panchayat level.

(5) At each such level, there should be a team-leader or coordinator of the team in over-all charge of the various programmes and projects. He will have to establish a well-defined relationship with the appropriate level of supervisory authority or the next higher local popular authority as the case may be.

(6) The role of the parent department may be restricted to cadre management and providing necessary technical support or guidance, enforcement of rules or regulations or administration of state-run institutions and services, and control of their budget as well as inspection and monitoring etc. of the local area programmes. The present-day nearest working model is perhaps the working of the cooperative department vis-a-vis various cooperative institutions.

(7) In a way, the break-up of the hegemony of the technical and development departments will help and promote a better and more congenial climate for the emergence of area-wise, regional sub-regional, district or lower field-level units of development manned by multi-disciplinary teams of experts and technical personnel under their control.

It is not intended to stipulate the functions or the organisational and staff composition of different levels of operations, except to stress that in the existing set-up of Rajasthan, or as per our recent traditions, there is as yet no clear-cut picture of team role of a regional or range-level coordinating and supervisory team except perhaps in the recent models of the commissioners for command areas, or the desert development and tribal area development commissioners and the like. Similar set-ups under regional development commissioners may be continued, rather strengthened.

XIV

AREA-WISE BUDGETS

Perhaps the biggest innovation which is required to be introduced would be to provide budget provisions not for the departments as such, but for the various watershed regions and other viable and distinct administrative-cum-development units in respect of the transferred or delegated functions and services. The regional or rural level heads of the local technical agencies should have the right to suggest transfer of funds from one activity to the other, depending upon the local requirements. At least, in respect of the agricultural and allied sectors including the programmes for rural development and watershed management in particular, the budgets should be drawn up on "a local area to local area" or "region to region" or "watershed to watershed" basis, instead of the present device of breaking up the departmental budgets for the district, sub-district or the special areas.

Table 1

Area, Irrigated Area, Production and Yields of Major Crops for the Years 1952-53, 1966-67 & 1978-79

Area in '000 Hectares Production in '000 Tonnes Yield in Kg/Hectares

S.No.	Crops	1952-53	1966-67	1978-79
KHARIF				
(1)	*Rice*			
	Area	71	94	211
	Irrigated Area	N.A.	N.A.	80
	Production	46	24	234
	Yield	639	275	1,109
(2)	*Bajra*			
	Area	2,755	5,115	4,535
	Irrigated Area	N.A.	N.A.	29
	Production	542	1229	1147
	Yield	196	240	253
(3)	*Pulses*			
	Area	1,268	1,950	1,762
	Irrigated Area	—	—	2
	Production	140	257	257
	Yield	110	130	146

Contd.

S.No.	Crops	1952-53	1966-67	1978-79
(4)	*Oil Seeds (Kharif)*			
	Sesamum			
	Area	348	728	423
	Irrigated Area	—	—	—
	Production	67	63	75
	Yield	192	87	177
(5)	*Groundnut*			
	Area	29	199	384
	Irrigated Area	—	—	9
	Production	13	76	239
	Yield	448	381	617
RABI				
(6)	*Wheat*			
	Area	741	961	191
	Irrigated Area	—	—	1,544
	Production	793	872	2874
	Yield	1,075	907	1,443
(7)	*Barley*			
	Area	244	439	406
	Irrigated Area	—	—	313
	Production	580	448	546
	Yield	1,334	1,084	1,345

Contd.

S.No.	Crops	1952-53	1966-67	1978-79
(8)	*Gram*			
	Area	568	1117	1748
	Irrigated Area	—	—	306
	Production	307	518	1589
	Yield	538	464	909
(9)	*Pulses (Rabi)*			
	Area	6	14	47
	Irrigated Area	—	—	16
	Production	3	6	29
	Yield	431	411	617
(10)	*Oil Seeds (Rabi)*			
	Area	167	303	419
	Irrigated Area	—	—	187
	Production	55	62	243
	Yield	1014	682	580
COMMERCIAL CROPS				
(11)	*Sugarcane (Kharif Irrigated)*			
	Area			
	Irrigated Area	19	32	60
	Production	414	393	54
	Yield	22246	12190	36600
(12)	*Cotton (Kharif Irrigated)*			
	Area	145	250	403
	Irrigated Area	—	—	346
	Production	103	184	570
	Yield	124	133	240

Table 2

Situation	Years	India	Rajasthan	Percentage of All India
1. Total Cropped Area (lakh ha.)		1,723.11	174.96	10.15
2. Gross Irrigated Area (lakh ha.)		459.10	34.51	7.51
3. Area under Principal Crops		(Area in lakh ha./Production in lakh tonnes/Cotton in bales of 170 kg.)		
BAJRA	Area	105.95	42.62	40.23
	Prod.	40.34	3.80	9.42
MAIZE	Area	57.52	8.79	15.28
	Prod.	55.78	5.74	10.29
JOWAR	Area	164.49	8.55	5.20
	Prod.	113.20	1.53	1.35
KHARIF PULSES	Area	72.31	15.34	21.21
	Prod.	15.70	0.62	3.94
GROUNDNUTS	Area	72.38	2.92	4.03
	Prod.	57.72	0.71	1.23
COTTON	Area	80.78	3.87	4.79
	Prod.	76.98	4.79	6.22
WHEAT	Area	219.64	20.69	9.42
	Prod.	315.64	26.96	8.54
BARLEY	Area	17.45	4.21	24.13
	Prod.	16.16	4.85	30.01
GRAM	Area	68.28	13.48	19.74
	Prod.	32.80	7.31	22.29
SESAMUM	Area	23.84	3.17	13.30
	Prod.	3.71	0.12	3.23
MUSTARD	Area	34.75	3.66	10.53
	Prod.	14.33	1.64	11.44

Source : Agricultural Statistics of Rajasthan compiled by Directorate of Agriculture, Jaipur (mid-seventies)

Table 3

Livestock Population in Rajasthan during the Years 1951-72 and 1977

	1951	1972	1977
1.0 Cattle	**107.82**	**124.70**	**128.96**
1.1 Male over 3 years	31.82	39.71	40.08
1.2 Female over 3 years	40.00	46.12	48.62
1.3 Young Stock	35.94	39.17	40.26
2.0 Buffaloes	**30.45**	**45.93**	**50.72**
2.1 Male over 3 years	1.52	1.47	1.90
2.2 Female over 3 years	15.97	24.92	27.85
2.3 Young Stock	12.96	19.54	20.97
3.0 Sheep	**53.87**	**85.56**	**99.38**
4.0 Goats	**55.62**	**121.61**	**130.07**
5.0 Horses & Ponies	**1.27**	**0.48**	**0.49**
6.0 Camels	**3.41**	**7.45**	**N.A.**
TOTAL	**252.43**	**389.73**	**409.63**

16

Area Development Authorities Conceptualised: Designing Area Development Programmes in Agricultural and Allied Sectors

NATHU RAM MIRDHA

The concept of 'area development' although not alltogether new in India has rarely been used in the same sense. Sometimes it is applied in the context of an integrated development of a whole region. At other times, it is applied in relation to the development of a very small area of a few hundred hectares. Area development in the wider sense embraces both agricultural development and rural industrialisation as well as development of economic and social infrastructure. Broadly speaking, when applied to agricultural sector, area development envisages the planned utilisation of all agricultural resources in a given area. The unit for planning integrated area development would then comprise an area which has a substantial measure of homogeneity in terms of land and water resource situation. Such a homogeneity would ensure that similar types of programmes are undertaken in a proper time sequence in order to exploit the development potential of the area to the maximum extent possible. Understood in this manner, it is possible to think of a variety of field situations which would admit different modes of development depending on the resource endowment of the different areas. For example, there could be area development programmes for commanded areas of irrigation projects, dry farming areas, drought-prone areas, desert areas, hill areas, tribal areas etc. In each of these areas, the types of programmes to be taken up for implementation will naturally be different. For integrated area development, it is not enough to piece together different programmes operating in a given area. It is imperative that the initial project formulations must be on a comprehensive basis and must provide for the fullest development of the resources of the area in order that the bulk of the people residing in the area could partake in the benefits of development. For instance, in an area served by canal irrigation, the important elements of an area development programme would be (i) development of land and water resources; (ii) consolidation of holdings; (iii) crop planning and agricultural extension; (iv) development of infrastructure; and (v) creation of an organisational and administrative framework for development. The objective should thus be to re-structure the production base of an area in terms of its land and water resources by undertaking measures such as land shaping, re-alignment of water courses, construction of the drainage system and strengthening of other activities which are complementary to these programmes.

At present cultivation in dry farming areas is characterised by both poor return and greatest instability due to uncertain rainfall and lack of irrigation facilities. Integrated development of dry land agriculture is important both for economic and social reasons. Accordingly an integrated package of technology involving the simultaneous application of all the practices developed by research scientists is being applied in selected areas in order to create a better impact on productivity and income in dry areas.

The programme for the development of drought-prone areas envisages creation of facilities which are physically and technologically possible. An area development project is being advocated which will include survey of resource endowment, in particular, availability of surface and groundwater, laying down appropriate farming techniques and choice of suitable drought resistant crops which alone can save these areas from the adverse effects of fluctuations in rainfall. In many of these areas which are hilly and undulating, the main problem is not of a low rainfall but of timely rainfall and the surface run-off of water which makes cultivation difficult. Here the principal need is to conserve water through farm level and village level schemes so that protective irrigation becomes possible.

The desert areas are the most backward in our country. Doubtful rainfall, sand storms, sub-soil salinity and alkalinity in places and nomadic livelihood based on cattle and sheep have made these areas difficult for development. In these areas, we have to think seriously in terms of providing adequate water wherever possible by taking water from remote catchment areas. The Rajasthan Canal scheme and Haryana Lift Irrigation scheme will solve the problem in large parts of Rajasthan and almost the whole of Haryana desert. In Gujarat and Rajasthan (Jalore and Barmer districts) the Narmada waters will ultimately transform about 10 lac acres, i.e., one million acres of desert into areas of crop production through irrigation. The desert development programme now contemplated is intended to build up vegetation in the deserts so as to push back the encroachment of the deserts on cultivated lands. Recently, a close look at the programme has revealed that there are very good possibilities of animal husbandry and forestry growth in these programmes.

In the hill areas where people are practicing subsistence agriculture, there are immense possibilities of introducing horticulture which will pay off much better than the cereal routine now being adopted by the hill area people. Similarly, sheep rearing can be a profitable occupation in the hills. Obviously, a change-over to horticulture and animal husbandry will require the support of necessary infrastructure in the shape of improved communications, storage, marketing etc.

In tribal areas, the practice of 'jhum' cultivation is causing immense damage to the vegetation in the hills and creating problems of soil erosion. Well thought-out programmes have to be launched in these areas for the development of tribal agriculture. Controlled 'jhum' with contour bunding and rotation of area, recommended and accepted more than a decade ago, will help revive soil fertility in areas subject to shifting cultivation. The pilot experiment in Vishakapatnam hill areas where large areas of badly damaged 'Podu' lands have been reserved has shown how such areas can be transformed by suitable terracing and cultivation into lucrative coffee plantations.

Agricultural development in the way it was proceeded during the last decade or so is still confined to certain to certain pockets of prosperity while large parts of the country continue to remain as backward areas in agriculture. This has resulted in imbalances among regions and the people. If this is not corrected in time, it will ultimately impede the very process of development in the country. Although the approach to our planning has always recognised the need for achieving inter-regional balances in development, there has been a marked tendency for economic disparities to get accentuated as between regions. Our planning should facilitate the development of each region according to its potentialities so as to make the optimum use of the resource endowment of the area. In industrial development also, the resource availability will determine the nature of development. Priorities for different aspects of development has therefore to be determined for different states and regions according to their needs and potentialities. In order that such balanced regional development is possible, it will be advisable to conduct proper studies, identify the resource potentialities in different areas and divide the country into various zones for planning and development purposes.

For integrated area development, it is necessary to design area development programmes in such a manner that the programmes per se reinforce and supplement each other in the overall economic development of the area. Apart from infrastructural services and inputs necessary for development in agriculture and allied industries, choice of suitable cropping pattern has to be made keeping in view the agro-climatic and soil conditions based on prior research. Inputs like improved seed, fertilisers and pesticides and credit have also to be provided. Bullock power and agricultural machinery has to be arranged. Till recently, very little work was done in this field in India although several multi-purpose irrigation projects have been undertaken. Recently, ayacut development programmes have been initiated in several states which emphasise the need for a coordinated approach in respect of command area development.

Rajasthan Canal Project

Rajasthan Canal now under construction ranks among the biggest in the world. On completion the canal will irrigate a total cultural area of 12.62 lakh hectares and a gross commanded area of 30 lakh hectares in Rajasthan. The area forms part of the great western desert and lies in Ganganagar, Bikaner and Jaisalmer districts. Most of the area is at present unsettled and uninhabited arid desert. Rainfall is scarce and uncertain being about 3 to 10 inches per year. Most of the area has deep sandy soils which are permeable, draughty and subject to wind erosion. A portion of the flat patches is of saline alkali soils. A high level committee comprising experts from Central and state governments has already evolved optimum cropping pattern for the area. Studies conducted by the UNDP project revealed the problem of levelling of the dunes in 8 lakh acres. There is need for proper soil and water management in the command area i.e. land levelling, water courses and field channels, lining of water courses, afforestation programmes etc.

Chambal Project

Chambal irrigation project is one of the major irrigation and power projects completed during the last two decades. The project consists of three reservoirs, namely Gandhi Sagar, Rana Pratap Sagar and Jawahar Sagar, and one barrage at Kota on river Chambal. It has been designed to provide water for irrigation in an area of 5.67 lakh hectares in the states of Rajasthan and Madhya Pradsh on equal basis. The command area of this project lies in Kota and Bundi districts of Rajasthan. The climate of the area is typically monsoonal with a long dry season and a short rainy season. The area is drained by system of nallahs. After the introduction of canal irrigation, these remain perennially wet with consequent aquatic wet growth and increased soil erosion. This has restricted the channels obstructing drainage raising ground water levels and increasing salinity problem. The canals were becoming infested with submerged weeds and within a few years the carrying capacity of the canals had been reduced to less than 50 per cent. The canal is being closed during May and June for clearance of weeds by means of both mechanical and biological measures.

These problems have not only adversely affected utilisation of irrigation potential but also resulted in low increase in agricultural production. In this background a UNDP aided project for land and water management and development in the nature of a demonstration was initiated in Kota district. The project took up integrated development of land and water resources of a compact area in the village Digod. However, for the entire area there is need for land levelling, construction of lined channels and field drainage for efficient and speedy use of the potential created and to prevent waterlogging. Proper regulation of water supplies and irrigation are also needed in the command area.

Each of the above areas has very great potential for agricultural development and absence of requisite measures of development will result in non-fulfilment of expectations of irrigation and expectations of growth of crops. Integrated development of command areas is also vital for full exploitation of the investments made in creating large irrigation potential.

As already emphasised, development of infrastructural facilities and services in a given area should be so planned that it should induce rapid growth of the area in a self-sustaining manner. The building up of an efficient infrastructure should, therefore, either precede or at least go hand in hand with the designing of an area development programme. To illustrate, provision of marketing, storage and communications is an essential component of the area development programme. Good marketing arrangements are a necessary pre-condition for profitable agriculture. Building up this infrastructure through normal course will take a long time. Meanwhile, lack of marketing facilities can lead to low prices and thereby hinder agricultural progress. There is need for developing regulated markets which include auction floors, auction sheds, storage sheds and provision of ancillary facilities like driers etc. It also requires all-weather roads from the producing areas to the marketing centres.

In designing an area development programme particularly in the command of a new irrigation project, there are a number of related aspects involving location of new village habitations which are required to be planned and executed. It is noticed that normally the villagers tend to stay near their fields instead of moving on the new settlements. Their resistance is due to the absence of proper living conditions in the new habitations. The farmers have to be attracted to the new settlements by providing basic amenities like good houses, primary schools, primary health centres, panchayat ghars, cooperative stores, etc. This is essential both for enabling the farmers to have a wholesome community life and also for a proper utilisation of irrigation facilities.

Executing an integrated area development programme on the lines described above, would require manpower of different kinds drawn from different state departments and organisations requiring the fullest possible coordination would be beyond the competence of a single state department of organisation. Development of an area of the size and the extent of Rajasthan Canal or the Chambal command requires a well planned and coordinated action of different departments like agriculture, irrigation, animal husbandry, cooperation, community development, P.W.D., Finance and various other financial institutions like ARC, land development banks, commercial banks etc. It would, therefore, appear necessary to create a separate area development authority or agency to be in overall charge of executing the area development programmes. A corporate authority having adequate expertise drawn from various disciplines should be able to undertake such a task.

17

Planning Priorities and Development Policies in Agriculture

BHUPENDRA HOOJA

THE PHILOSOPHY OF DEVELOPMENT IN INDIA

We are now almost half way through the Sixth Five-Year Plan and the planning era is 30 and odd years old. Perhaps it is time to look back at the starting line and re-examine or re-state the goals and perspectives of our all-out national effort on many fronts and in many sectors of development so as to create a more just and equitable social and economic order and to give new opportunities for growth and self-expression to the teeming millions.

The over-riding problems of the teeming millions during the last couple of centuries, particularly under the British rule, have been of mass poverty, hunger and lack of economic opportunities in a situation of 'arrested growth' of economic and social stagnation largely due to the country having been drawn into the colonial or imperialist system of exploitation. Consequently, the challenges before the national movement in the freedom struggle and ever since have been to overcome these constraints. Naturally, the success or failure of the national efforts needs to be viewed against this background; perhaps not by the usual measurements of economic growth, per capita income and such other econometric yardsticks, but by the quality of life which is much more than material wealth and physical well being of the average citizens.

Almost hundred years ago, in 1876, Dadabhai Nairoji had written a paper on “The Poverty of India”. Since then, as intelligent public opinion and the national movement for reforms and then for freedom grew, its social, political, and economic objectives began to be more and more clearly defined. Gandhiji came out with his dream of an India, as a land where there would be no poor nor any rich, his ideal of *Gram Swarjaya* and his prescription of constructive work, self-help, *charkha* and basic education to bring back the old days of self-reliant and autonomous village communities. The message of the Indian National Congress spread far and wide through the tireless work of thousands of *satyagrahis*, reformers and constructive workers and the successive movements launched by it brought an awakening in the rural community and the agrarian situation came to the fore. As Nehru has conceded, (in his book *The Discovery of India* and elsewhere) a whole generation of political workers, social thinkers and enlightened public-men learnt their economics by going out to the villages and coming face to face with the realities there. In 1931, the Congress adopted a comprehensive economic programme, which became more rural or agrarian in content by 1936. Meanwhile, the political creed had crystallised into the call for complete independence; and with the partial introduction of a federal system and provincial autonomy in the provinces, the Congress had got ready to fight the elections to the new (provincial) assemblies, as a result of which Congress ministries were able to assume office in seven provinces, while workable coalitions were established in others.

Pioneering Work of NPC

In this changing situation came the first-ever plan-frame of future economic development and a clear enunciation of social objectives and policies. The National Planning Committee or NPC set out some broad objectives for improvements in food, clothing and housing standards, increased production in agriculture and industrialisation, employment opportunities, increase in per capita income, extension of public utility services and educational as well as medical services. All these were spelt out as steps towards the goal of “national self-sufficiency” so as to break away from the imperialist economic system, to promote long-term and sustained growth, for which some institutional and structural changes were also spelt out, and thus to "meet the domestic needs to food, raw materials and manufactured goods".

Since neither the problems of poverty and unemployment could be faced nor economic regeneration ensured without large scale industrialisation, the NPC was in favour of the development of heavy key industries, medium-scale industries and cottage industries in a rather comprehensive but flexible or varied institutional set-up, without at the same time ignoring agriculture, which was the mainstay of the people. The NPC plan had a ten years' time frame.

Since, at that time, freedom was nowhere in sight and other international and national level happenings overtook and interrupted its work connected with the active association and cooperation of hundreds of the best brains from different walks of life, the goals and objectives set out and the strategies of the structural changes and institutional devices then considered proved very useful when the real opportunity came about a decade later. Infact NPC plan exercise was a maiden attempt to pinpoint our social objectives of an Indian brand of socialism.

In this context, Nehru's reflections may be quoted: "...Yet it became clear to me that our plan as it developed, was inevitably leading us towards establishing some of the fundamentals of the socialistic structure...(by)...limiting the acquisitive factors in the society ... (or by) ... planning for the benefit of common man ...and giving him opportunities of growth...in the context of democratic freedom and with a large measure of cooperation of some at least of the groups who were normally opposed to socialistic doctrines".

Socialistic Blueprint: Role of State in Industrialisation

The very essence of planning was a large measure of regulation and control. Thus, while free enterprise was not ruled out as such, its scope was severely restricted. In regard to defence industries, it was decided that these must be owned and controlled by the state. Regarding other key industries, the majority were of the opinion that these should be state owned, but a substantial minority of the committee considered that state control would be sufficient. Such control, however, of these industries had to be rigid. Public utilities, it was also decided, should be owned by some organ of the state, either the central government, provincial government or a local board...it was suggested that something of the nature of London Transport Board might control public utilities. Further, in regard to the agency in state-owned industries, it was suggested that as a general rule an autonomous public trust would be suitable. Such a trust would ensure public ownership and control would creep in under direct democratic control. Co-operative ownership and control were also suggested for industries.

Land Policy

Regarding the policy for management of land resources, the NPC had laid down the principle that "agricultural land, mines, quarries, rivers and forests are forms of national wealth, ownership and management of which must vest absolutely in the people of India collectively."

Nehru further commented and observed in this connection that "the co-operative principle should be applied to the exploitation of land by developing collective and co-operative farms. It was not proposed, however, to rule out peasant farming in small holdings... but no intermediaries of the type of talukadars, zamindars, etc. should be recognised after the transition period was over. Collective farms were to be started...on cultural waste land. Co-operative farming should be combined with individual or joint ownership."

"We or some of us at any rate hoped to evolve a socialised system of credit. If banks, insurance, etc. were not to be nationalised, they should least be under the control of the state thus leading to a state regulation of capital and control. It was also desirable to control export and import trade. By these varying means, a considerable measure of state control would be established in regard to land as well as industry as a whole...".

From NPC to the Constitution and Plan Documents

It must be realised that Nehru and his brains-trust had rejected the Soviet model of planning (as also the Soviet state structure) though he and his colleagues and followers have been often accused of having a bias towards Soviet Russia.

Among the controversies which have survived that period are the pet ones like private enterprise vs state sector, the primacy of agriculture over industrialisation, the role of small and cottage industries, and so on. But, if we read through the NPC exercise, or even Nehru's recollection and revaluation, we will be struck by the fact that theirs was a more or less balanced, rational and scientific approach of developing the resources through appropriate technical and development efforts for attaining the goals of the welfare of the mass of the people, and to restore to them the dignity and freedom of opportunity thèy deserved – all this within the framework of a system of democracy and decentralisation or de-concentration of political and economic power.

In any case, NPC had presented a broad national consensus. The different views of Communists, Socialists, Gandhites or Muslim League followers had yet to emerge on the scene. Having noted that, we may proceed to review the efforts of the post-independence generation and see where compromises or retreats have been made, where contradictions have developed and where we have failed or succeeded in approaching the NPC goals.

A parting word may also be added. Gandhiji himself did not have any occasion to reject or criticise the work of NPC until the end as it was mostly in line with his thinking.

SHOCKS ON THE EVE OF FREEDOM AND AFTER

India had to pass through quite a few traumatic experiences and shocks before planning could be initiated in right earnest and in a more congenial environment.

The Freedom Struggle: Quit India Resolution

One such experience was when the entire national leadership representing the mainstream of the freedom struggle and thousands of activists of the Indian National Congress were put behind the bars following the so-called Quit India Resolution of August 1942; but the struggle petered out or was subdued after a few months of initial (and in many parts of the country) violent or agitational reactions.

Bengal Famine

Then came the shock of wide-spread famine in Bengal and Orissa which has been often called a man-made famine, which overwhelmed both the official and popular minds, with millions dying of starvation and hunger and widespread signs of complete disruption of the rural economy and the logistics of transportation and break-down of official efforts to bring succour to the famine-striken people. Going through the last 100 and odd pages of Nehru's book, *Discovery of India* (as well as other contemporary literature and journalistic, academic or official reports) one cannot escape the conclusion that this indeed was a very stunning or shocking experience.

It led Nehru to write many pages on India's proverbial sickness, both in mind and body, of poverty perpetuated by the Raj and famine due to war. In his words, "while some people had prospered during the war, the burden on others had reached breaking point and as an awful reminder of this came famine, a famine of vast dimensions affecting Bengal, and East and South India." This catastrophe in a region "nearest to the theatre of war" not only injured the country's capacity for defence, but it also brought out the deficiencies of the political, administrative and economic systems, and both the good and bad qualities of the Indian people.

Partition: Some Economic Consequences

Another shock which nearly shook the very foundations of Indian polity and disrupted the economy, already over-strained or very near ravaged during the war, was the partition of the sub-continent which accompanied the grant of independence by the British. Apart from the emotional, psychological, social and cultural, political and administrative shocks, this fact broke as under the previously unified economic structure, particularly in the fields of agriculture, irrigation and rural development. The rich, fertile and irrigated parts of West Punjab, Sind and East Bengal went out of the Indian economic and political system to constitute the new state of Pakistan. From a state of near self-sufficiency, India was reduced to a deficit and dependent country.

Naturally, and it was obvious to all concerned policy-makers that above three unfortunate happenings or shocks following each other in quick succession, had completely disturbed the delicate equilibrium of food and other essential supplies One of the major concerns of the new Indian government, caused by the partition or the mass movement and 'exchange of populations, was to meet the food requirements of an ever-growing number of people in the new Indian state.

While on the political and ideological front, the Indian people and their leaders managed to give to themselves the new

Constitution enshrining all the values which they have cherished since the beginning of the freedom struggle, on the economic front and particularly in the agricultural sphere, they tried to patch up and bolster a crippled, broken, and falling apart economy. In fact, the First Five-Year Plan was more or less a patch-up affair of putting together various types of schemes or on-going programmes of great urgency, including the war-time grow more food campaign.

Jammu & Kashmir and Indo-Pak Conflict

However, the conflict with Pakistan tribesmen and defence forces in Jammu and Kashmir and the imminent threats of disintegration due to internal turmoil and external aggression also took considerable toll of the limited resources of the country and forced the authorities to divert their attention as well as efforts towards repairing and building up a reasonable defence system.

Meanwhile, there was a heavy drawal on the sterling balances accumulated during the war (at great sacrifice of the colonial people) and this advantage was also soon run down. By the time the Second Five-Year Plan was launched, the unity and security of India had been more or less assured and the immediate political, economic and social after-effects of partition and Pak aggression had also been brought under manageable control; and not only the Constitution had been adopted and princely states merged with the Union of India, but the first-ever democratic elections based on universal adult franchise had also been successfully conducted and a representative government established.

Population: Another Threat

However, there has been a constant drain on the resources of the country and its achievements from another internal factor. Right from the days of the National Planning Committee, national leaders and economists, planners, social scientists and sociologists have been deeply concerned with the serious problem posed by an unprecedented growth in population. Coming on top of the rather complicated and, in some respects, a hopeless post-war and post-independence situation of mass migrations, political, social and economic disturbances and the dislocation, if not disruption, of the established economic system and the pressures of restructuring a new political environment and institutions, this realisation about the growing population and its growing needs for food, clothing, educational and medical coverage and the like, became (and has always remained) one of the major concerns if not an obsession, with the authorities whose responsibility it was to guide and direct the destiny of this new born nation along proper lines.

THE ESSENCE OF PLANNING

First Steps in Agriculture

Thus the first step to build up the economic infrastructure through multi-purpose irrigation and power projects (the so-called new temples of India) was taken with a view to achieving higher levels of production and employment and better living standards.

Land reforms were also introduced to change the basis of the old socio-economic order and to bring about a more equitable distribution of land, a major factor of production. Though contrasts and disparities have persisted (between marginal, small and medium holdings), this provided a great impetus to agricultural activities and investments.

The war-time programme of grow more food was also continued; rather it was stepped up. More and more cultivable areas were brought into use and more of such lands were covered with state-sponsored seasonal and perennial irrigation, as well as irrigation developed by the cultivators themselves. Improved varieties of seeds and other inputs were developed and provided through agricultural extension and C.D. blocks. Improved implements, better compost pits, improved arrangements for cooperative credit and locally suitable improved practices were introduced, duly supported by appropriate research activities and suitable incentives or subsidies.

Some Salient Facts of Agricultural Development

Since the beginning, the attempt has been to awaken and involve the rural masses, cultivators and cattle-breeders, artisans and craftsmen in the process of agricultural development. This has been considered necessary because nearly 80 per cent of the people of India have continued to live in the rural areas and are land-bound even today.

They live and work on their small and medium sized (at least not very large) holdings and between them manage to contribute nearly 50 per cent of the nation's total wealth. As per 1970-71 data, of nearly 70 million operational holdings, more than half were below 1 ha. each, and 24 million between 1 and 4 ha. each. But the total area of holdings of these categories 56 million ha. (13 m. ha. and 43 m. ha.) was about two-fifth of the total area (135 m. ha.) of operational holdings. A little more than 10 million holdings above 4 ha. covered about 80 million ha. area. Whatever the government plans or priorities, ultimately the success or failure of their implementation depends upon the decisions, resources and efforts of the country's countless people (78 million cultivators and 48 million agricultural labourers as per 1971 Census).

It may be worthwhile to point out that the situation in India is quite different from what it is (or may have been in the early stages of growth) in the advanced western countries like USA or in Latin America or anywhere in Europe, or in the Soviet Union or in China or Japan, the countries which have done well in the industrial or agricultural spheres. The differences are much too obvious to be compared or recounted. Suffice to say that just by the force of sheer number, India is the largest democracy. It is also the land of the second largest number of cultivators and perhaps the foremost as far as small-scale private farming is concerned. China with the largest rural or agricultural population, differs from India in the sense that there is a lot of collective and cooperative enterprise.

The disparities in holdings size and their areas mentioned above are also significant both from the sociological and the productive points of view, particularly the number of agricultural labourers and the small cultivators.

Looking back over the past over thirty years' progress, we can (and should) appreciate how tremendous the task has been to motivate, guide and encourage the scattered and often isolated farming communities in India and draw them into the multi-dimensional advance towards modernisation based on new science and technology.

During these years, the net sown area has increased from 119.40 million hectares in 1951-52 to 143.08 million hectares in 1978-79, an increase of more than 18 per cent, and the gross areas has increased from 133.23 to 173.30 million ha. or more than 28 per cent. During the same period, the net and gross irrigated areas have also increased from 21.05 and 23.19 million ha. to 39.09 and 48.48 million ha. respectively, an increase of about 50 per cent. Never before has such expansion or growth taken place in so short a period in any other country in the world.

Both these indicators suggest efforts on the part of the government as well as the cultivators in bringing more and more areas under cultivation and irrigation. If one can compute the approximate cost of such efforts, including the cost of land allotted or occupied and then evaluate all sorts of land preparation and farming operations, it may be appreciated that both the state and private (as well as institutional) investments have not been lacking. It is another matter that all cultivable lands may not have been prepared or developed well, or in respect of irrigation there may have been time-lags between the creation of potential and its proper or effective use after full or optimum development of command areas.

However, while the total areas under principal cereal crops has increased from 78.18. to 105.57 million ha. during the period under study, most of the increase has been in respect of fine grains like rice (from 29.83 m. ha. to 40.20 m. ha) and wheat (9.47 m ha. to 22.22 m. ha). In the case of other m1in crops such as jowar, bajra or maize, the increase in cropped areas has not been phenomenal.

The total area under principal foodgrains, including grams and pulses etc. has increased from 96.96 to 128.12 million ha This should also give us some idea of the costs of various inputs like the seeds, compost, fertilisers, pesticides and the use of various types of tools and implements and the sources of energy for agricultural purposes.

Data regarding the production, import and consumption of chemical fertilisers are also quite revealing and significant. In the year 1951-52, only 58,700 tonnes of nitrogenous (N), 6,900 of phosphatic (P_2O_5) and hardly any potasssic (K_2O) fertilisers were consumed, out of which indigenous production was 28,900 tonnes of N and 9,800 tonnes of P_2O_5. But almost 25 years later, in 1979-80, the consumption of N had gone up to 34,99,000 tonnes with imports being slightly more than one-third. In the case of P_2O_5, consumption had increased to 11,50,300 tonnes, while imports were less than one-fifth. In 1979-80, the K_2O consumption was 6,06,500 tonnes, and almost three-fourth of this or 4,73,200 tonnes was imported as there was no local production. However, in spite of substantial production within the country, the total value of imported fertilisers was to the order of 5,545 million rupees. It is also worth mentioning that imported fertilisers have generally been subsidised by the government, besides other subsidies for small or marginal or scheduled castes and scheduled tribes farmers.

The total production of foodgrains in the year 1978-79, for which the respective totals of areas under different crops have been mentioned earlier, came to about 131.37 million tonnes. Accordingly, the yield per hectare was 1025 kg. Major crops were rice 53.3 million tonnes, wheat 34.98 million tonnes, jowar 11.56 million tonnes, followed by bajra 5.52 million tonnes, ragi 3.04 million tonnes, small millets 1.93 million tonnes, barley 2.31 million tonnes, grams 5.84 million tonnes, tur (pulses) 1.91 million tonnes and other pulses 4.42 million tonnes.

Diversity and Set Patterns

The area under various crops and the production as well as the average yields per hectare differ from state to state or region to region according to climatic conditions and according to different types of infrastructural facilities available such as irrigation or electrification and the like.

What we often say in sociocultural and political terms about the unity of India lying in its diversity applies equally well in the case of agricultural production. Because of the diversity of soils and climates, there is a wide variation rather a variety of principal crops as well as a diversity of local cropping practices and techniques. But within a particular state or region, the cultivators are not left with much of an option. They do not have many alternatives to choose from. Though six principal crops can be grown during the kharif season and almost a similar number during the rabi, it is mostly the weather factor which determines whether the cultivators would have a particular set or combination of crops or the other. For example, if the rains fail or the irrigation facilities during the kharif are interrupted or restricted, it is quite possible that most rice growing cultivators of a particular area, say north-western India or north-western Rajasthan may decide to go in for wheat and gram or barley during the rabi. Similarly, during the kharif season itself, they may opt for jowar, bajra and maize, whichever has been the main traditional crop. In case abundant irrigation water is available or rainfall is timely and at frequent intervals, they may choose to go in for commercial crops and even have very encouraging production levels. But they will still have their preference for dry crops. However, it has been noticed that, particularly during the kharif season, cultivators do not utilise irrigation and other inputs to the maximum extent on crops which have comparatively less commercial value.

Fluctuations and Deficits

It is largely because of the seasonal fluctuations of rainfall that the production levels keep on varying from year to year. Of course, because of the emphasis on rice and wheat under the "green revolution" and because of the assured and gradually increasing support price, there is a greater premium on these two crops as on some other commercial crops. During the last few years, it has been increasing felt that if similar attention and support with suitable subsidies or incentives, etc. would be forthcoming, the production of various other crops can also be stepped up and stabilised.

The inherent limitations of our agricultural sector can best be appreciated from the recent estimates and press news that kharif production during 1982 has been hit by erratic monsoon, with deficient rainfall in Jammu and Kashmir, Tamil Nadu, and large parts of Rajasthan, Gujarat, Andhra Pradesh and Karnataka. U. P., Bihar, West Bengal and Orissa have also suffered either from late rains and long dry spells or floods or both. For the country as a whole, this will be the second year in succession of erratic monsoons; but for states like Rajasthan, the trouble has been worse and longer.

How such erratic monsoons can change the food situation may be seen from a recent review by the National Council of Applied Economic Research that while in a good or normal rainfall year (1978-79) the production of kharif crops had been 78 million tonnes, the next year (1979-80) it declined to 63.3 million tonnes. Thus, the challenge of a stable or assured production of kharif crops remains.

Last two years' bumper crops have also been gifts of good monsoons.

Towards a Green Revolution

Whatever the critics may say, and also those who have opposed the emphasis on industrial development and urbanisation and concentration of political and economic power, agricultural development mainly keyed towards the production of more foodgrains as well as commercial crops has remained one of the main planks of the programme of planning all through these years.

As far as plan funds are concerned, there were huge investments in irrigation projects and other infrastructural facilities. But, in a country where agriculture is the main economic activity of millions of small and scattered cultivators, it would not be fair to measure the government intentions or contributions by investments only. One has to add up scores of activities or schemes in other areas like research, extension and development, marketing legislative and even regulatory spheres to have a total view or assessment of what was intended or what could be achieved.

But what has been achieved in Punjab and Haryana or elsewhere, where benefits of assured irrigation were available would not have been possible but for land reforms or institutional changes, the dynamism of the people and their administrations, and some new infrastructure facilities like power and irrigation, cooperative and commercial bank credit, storage and processing as well as network of roads, markets and the like. If we add up the real cost of all these activities, whether financed out of state resources or through co-operative channels or later on by nationalised commercial banks or by the cultivators themselves, it should not be difficult to appreciate that, on the whole, agriculture has not been a neglected sector as is often alleged.

Naturally, the strategy for agricultural development has changed from plan period to plan period and region to region. Sometimes when the weather and rain conditions have been favourable, there have been real spurts in the level of production of all types of crops, particularly the cereals, oilseeds and pulses and some other commercial crops as well as during 1983-84.

An Initiative of Millions

The critical comment that investments of plan funds on agriculture has never exceeded 21 to 25 per cent of all state investments is not quite valid.

When we discuss the agriculture sector (in its wider concept and connotation, including animal husbandry, cooperatives and such other sub-sectors), we should remember that this is a sphere of economic activity left entirely or mainly to the initiative of private enterprise of millions of small, medium and large cultivators who can tap resources from many quarters. Secondly, in large numbers, and particularly in the erstwhile princely states, the cultivators have been benefitted with tenancy rights and have been brought into direct relationship with the state as the principal land-holder in the country (on behalf of the people of India) in their capacity as tenants and they have also gained from various regulatory measures including revenue, forest, irrigation laws and other measures like low land revenue rates, fixity of tenure, consolidation of holdings and elimination of middlemen, both in respect of cultivatory practices, and marketing or processing activities.

An Overview of the Planning Era

Generally, the period of first 15 and odd years of planning is considered to have been one of recovery and reconstruction of the economy after the 1939-45 war and the disruption caused during the last phase of the freedom struggle. Various sectors of economy appeared to be developing at a good pace, while at the same time, foundations were being laid to build up a modern democratic state structure and a modern industrialised society. Besides agricultural and industrial development, many social and economic reforms were also introduced. Fortunately, this period remained, on the whole, full of hope and buoyancy. There were many favourable monsoons. The people at large seemed to be on the march, and the early results of some plan projects were encouraging so that the aftermath of the 1962 Chinese agression and the 1965 war with Pakistan could also be overcome to some extent, though there were disappointments too. And because of the growing defence needs, a serious lack of finances for planned development also began to be experienced, so that successive plans had to be pruned in the 'core' or the very hard core sectors. Besides shortages of food grains and consumer goods, inflationary pressures were also becoming noticeable. This is how "plan holiday" or a series of Annual Plans germinated in our country.

A CRISIS HAS SET IN

However, from late 1960s and early 1970s, a different scenario seemed to develop. By this time, certain contradictions in the strategy of planning and their consequent economic constraints had also began to show. For example, the stress on basic and key industries in industrialisation and the adoption of a capital intensive strategy of investments for development purposes, though these added to the productive assets of the nation, did not correspondingly result in more or adequate number of jobs, nor higher production in diverse fields on a sustained basis nor to increase in per capita or gross national incomes. On the other hand, and looking back at that period, it is not very certain whether any other alternative strategy such as a labour intensive programme of development, or comparatively low emphasis on industrialisation would have worked any better or more satisfactorily.

Shortcomings and failings were also witnessed in certain other respects such as in respect of land reforms which could not be successfully carried out to their logical end, or in dividing the maximum benefits from irrigation and power projects or in the production of items of daily consumption, even foodgrains, for which there was no escape from import. Nor was there sufficient training and development of managerial and technical skills so as to successfully man the various industrial projects, cooperative ventures or development departments. Above all, the need of "self-restraint in consumption" and "economy in expenditure" so as to boost savings could not be translated into a workable and popular proposition.

One of the objectives of planning being to provide equitable justice and basic rights, the contrast between the lifestyle and culture of the "haves" and "have-nots" could also not be diminished. Rather the gap between the rich and the poor has begun to widen; and strong groups and lobbies of the well-to-do and well-organised or more vocal lobbies have begun to assert themselves in many spheres. They seem to corner away most of the benefits of development including civic amenities, welfare facilities and incentives. In fact, (perhaps because of their clamour) there has been increasingly more and more emphasis on the provision of welfare amenities and social services, even when very serious constraints on resources are being felt. Such were the demands of the situation that in a system of democratic planning and development administration, and in a welfare state, necessary discipline or climate of abstinence in this regard could not be enforced.

Never was there a greater need for the restraining voice of Gandhiji and his philosophy or creed of self-denial or self-sacrifice than at this stage. There could have been no other better occasion than this to think of the egalitarian or socialistic views of Nehru.

But the pressures of planning seemed to have pushed these guidelines into the background.

Centralised Planning in a Democratic Set-up

Perhaps, to some extent, these problems and shortcomings were also due to the fact that we had deviated somewhat from the path originally set before us. We also sought to establish and work a system or structure of centralised (or centrally controlled and directed) planning on a very vast and large scale covering almost all spheres of human activity, economic social or cultural, under a democratic set-up and through democratic processes. But the status and the role of the privileged and the powerful had not been changed. That was also one of the compromises made in the blue-print of the NPC. The result was that there was a vast and unbridged gap between the policy approach or the implementation machinery and the scope of development activities of the government on the one hand and the expectations of the masses of the people from the process of planning on the other, and what the organised and powerful sections could corner. Perhaps because of the constraints of 'democratic planning', it was not possible for the government to effectively enforce discipline, to mobilise savings or to raise additional resources through taxation and other measures, or to enforce economy measures or to reduce consumption or to enforce its control or regulations or self-restraint on the defaulters and the deviators or amongst the people at large.

Ever increasing expectations of the people have had to be met willy-nilly, and at the same time, there has been the impact of external factors like threats of aggression, and the overall world-wide climate of accelerated armaments; and of a series of natural calamities and certain internal factors like domestic economic set-backs caused by the pressure of growing population, climate of shortages of goods and scarcity of foodgrains and other essentials emerging on the horizon. For the poorer and weaker sections, the problem has become all the more acute because of the vocal, organised and powerful sections of society clamouring for and claiming whatever they can of the national cake by way of more concessions and incentives and better terms and more easy living conditions for themselves. They can also overcome or escape the net of various government laws and regulations meant to curb such anti-social and anti-people trends. Two other phenomena which emerged during the period (the latter period of planning) are also worth mentioning. One has been that even though the state or the public sector undertakings have gradually come to assume "commanding heights in our economy", their working in a bureaucratic organisational framework and their accumulating loss notwithstanding, in the case of private sector, there has developed a growing tendency to shift attention from the manufacture of consumer goods towards the more profitable production of capital goods, thereby further accentuating shortages in the market.

Emerging Intermediate Classes

Capital intensive investment, both in the state and the private sectors, because of their reliance on more sophisticated and modern machines, has failed to create sufficient number of jobs for unskilled or semi-skilled persons at the lower (or base) level of society, while on the other hand, the managerial and intermediate sections of society as also traders, shop-keepers, money-lenders, professionals like doctors, lawyers, technocrats and a host of others have come to assume a dominating role in the social, economic and political spheres. This rise of the intermediate class has also been witnessed in many other countries emerging from their colonial past and drivern towards modernisation. The intermediate class has also got its counterparts in the rural areas, such as sections of large and middle-class farmers or the erstwhile feudal elements and peasent proprietors or those middle men, traders, etc. who have advanced ahead of their contemporaries in rural society in the process of modernisation. They have got the best of resources and facilities at their disposal in agriculture, mechanisation or electrification, availability and use of irrigation facilities and the application of sophisticated and costly inputs for production.

Today, members of this so-called intermediate class seem to dominate the political scene both in the urban as well as the rural areas, perhaps more in the rural areas. They also manage to by-pass some of the regulatory or controlling measures of the government and over-ride the interests of the comparatively poorer or weaker sections of society. Not only that, this class has obviously thrown in its weight with the industrialists and financiers and with the manipulators of the economic and political levers of power.

Much of what has happened in our country during the last ten years or so in the political, economic, or social and cultural spheres, can be understood better if we keep in mind that over the years, there has been a slow but steady concentration of economic and political power in fewer hands, which is sometimes shared with lobbies and strong organised groups of the intermediate class.

This is a distortion which has come about through the process of planning, which goes contrary to the basic policy objectives which were set forth in very clear terms both during the freedom struggle as well as in the framework of the National Planning Committee's reports and other documents of national importance like the Constitution or the various plan documents.

It seems that a sort of permanent crisis has set in, reflecting shortcomings and setbacks in many sectors or spheres of activity. Basically, it is a crisis of systems, rather of the transition from the old colonial order to a new equalitarian society, the outlines of which have been drawn in the objectives of the Constitution and the plans. There is no denying that much progress has been achieved in the industrial sphere, in agriculture and in social services. Still there is a lot to do. Many setbacks have been there. At the same time, quite a few bastions of power and privilege have been humbled. Productivity capacity is there, though all of it is not fully utilised. Food production has gone up; but the mouths to feed have also increased. Poverty because of low production has been reduced; but poverty due to mal-distribution or mal-functioning of the productive forces exists.

Yet, through all these recent ups and downs and setbacks, one fact which stands out is that neither the very rich class of industrialists or financiers, nor the intermediate class of professionals as mentioned earlier and well-to-do peasant-proprietors in the rural areas have had any setback or disaster. But the position of others is bad as the lower middle and working classes in the towns and cities and small or poor farmers and other weaker sections in the rural areas have been increasingly feeling the pinch of shortages in production or non-availability of consumption articles at reasonable prices or even necessary inputs for production in the market.

WHITHER AGRICULTURAL SECTOR

In the present situation, we have still to find a way out of the prevailing contradictions. It is ironical that with all the 30 years of plan efforts and shifts in strategies and tactics and with all the investments, particularly in the power and irrigation sectors or in respect of agricultural research, local production or import and distribution of inputs and various other agrarian reforms, we have not yet been able to reach a situation when sustained growth in the agricultural or rural sector may be possible. Now more and more economists have begun to realise and protest that the level of production has almost reached a "plateau" or a stage where it is more or less level or fluctuates between certain limits without showing any tendency for further or a sudden spurt. On the other hand, there is no stopping the increase in population.

The following extensive extracts from a recent writing by a contemporary economist are quite meaningful:

"Foodgrain production in the country reached the peak of 131.9 million tonnes in 1978-79. It was 109.7 million tonnes in 1979-80, 129.87 million tonnes in 1980-81 and 131.2 million tonnes in 1981-82. Considering the damage done to the kharif crops by the drought, production in 1982-83 is expected to fall by 7 to 8 million tonnes compared to the previous year."

"The stagnation in agricultural output cannot be explained by the technological factor or by the inadequacy of public sector investments in the farm sector. The area under irrigation and under high yielding varieties as well as the use of chemical fertilisers have all continued to grow, albeit a little slowly than before, over the past four years."

After quoting relevant data about the increasing use of H_1V seeds or fertilizers, the writer continues:

"The total cropped area, however, has not shown any increase since 1978.79 when it stood at 175.18 million hectares. It fell to 167.20 million hectares in 1979-P.O; but rose to 173 million hectares in 1980-81. Estimates for 1981-82 put it at 195 million hectares, which is slightly below the 1978-79 figure."

"Thus, if the total cropped area as well as production remain constant while the volume of inputs continues to grow, as indeed has been the case in this country for the past four years, the only conclusion is that output has ceased to be responsive to the extension of irrigation and an increase in the use of modern farm inputs."

"In other words, despite the increasing use of fertilisers and high-yielding variety of seeds, crop yields per hectare have ceased to grow. The only exceptions are the three cereals, namely, wheat, rice and maize, and one commercial crop, sugar-cane. Yields of all other crops show a decline."

"Andhra Pradesh, Assam, Kerala, Maharashtra and Karnataka showed varying degrees of improvement in yields in the case of rice; and Assam, Bihar, Gujarat, Maharashtra and Uttar Pradesh in the case of wheat. On the other hand, rice yields suffered substantial declines in Uttar Pradesh, Tamil Nadu and Rajasthan, apart from Punjab; and wheat yields in West Bengal, Karnataka and Himachal Pradesh."

"It appears that States like Punjab and Haryana, which first felt the impact of the Green Revolution in respect of wheat and rice, are fast reaching the upper limit of feasibility in yield increases. while progress in other States continues to be tardy."

"In the case of other foodgrains, particularly millets and pulses, the Green Revolution is yet to make an impact. The same is true for oilseeds, fribres and other commercial crops."

A number of questions have been raised by the writer, who has also tried to explain the recent trends thus:

"No single factor can explain the current crisis facing Indian agriculture. Among those which have contributed to the present stagnation in production and productivity are: (i) inadequate research in plant genetics for crops other than wheat and rice, (ii) the present structure of land-holdings and land tenures which prevents putting available land to the most efficient or economically optimum use, and (iii) the agricultural price policy which may be acting as a disincentive for the farmer to invest in land improvement and production."

He has further argued that:

"Little progress has been made in evolving high-yielding varieties of seeds for pluses, oilseeds and most commercial crops with the result that the area under these varieties in the case of these crops is negligible. Consequently, production has failed to respond to market stimuli because of the persistent low-yield levels of these crops."

"The second bottleneck in the way of increased agricultural production is the land system. Large land-holders are not putting their holdings to the most effective use. Yields on larger holdings are generally lower in West Bengal, Bihar, Orissa and eastern Uttar Pradesh. On the other hand, the size of holdings in the case of marginal farmers in particular and small farmers in general is so small that it inhibits the use of costly high yielding varieties of seeds and other inputs like chemical fertilisers and other pesticides."

"The third factor inhibiting growth of agriculture is the Centre's price policy. We are perhaps unwittingly preventing an increase in the production of wheat and rice by relating the procurement and support prices to the needs of the public distribution system. If production of cereals refuses to pick up beyond the levels already reached, it may be because there is no effective demand available to call for an increase in supply. We do not allow export of cereals even in bumper crop years because that would put pressure on domestic supplies and, therefore, prices. The procurement prices fixed by the government may be remunerative for farmers in areas where agricultural conditions are favourable, but they are certainly not so in the case of their less favourably placed brethren."

AN ENVIRONMENTAL VIEW

The other dimensions of the problem also need attention. As far as our land resources are concerned, these seem to have been tapped and harnessed to the utmost or maximum extent. In fact, forests and grass lands have begun to disappear and most lands are losing their vegetative cover. The poor productive capacity of various types of land in various regions seems to be under constant and heavy strain. Apart from the lands under cultivatory use or those being managed by the forest department and other agencies, there are vast tracts of land in different regions and agro-climatic zones of our country which are lying barren, uncultivated or unutilised, and many amongst these are gradually being eroded or getting degraded season by season because of neglect or lack of attention.

Though the planners and economists, and to some extent, some of the natural scientists, have not realised it yet, there are others mainly those connected with environmental sciences, who would like to see a more scientific and planned use of natural resources available in the country on a long term basis and who have also been warning against over-exploitation by way of cultivation, grazing or felling of trees and degradation of forests, and even by way of urbanisation or industrial use of otherwise fertile agricultural lands.

It is now being gradually realised that all planning and development activities must be based upon and closely linked with the nature and the quality of the natural resources available or land and water resources.

It must be appreciated that while hitherto our planning and development efforts in the agricultural sector have been based on certain pre-determined or calculated physical targets, or on the removal of socio-economic constraints or institutional changes, but now the time has come when such strategies or programmes should be modified or subjected to the inherent limitations or constraints of the nature or quality of the available natural resources.

Land Use According to Capacity

After a careful, detailed and countrywide survey and classification of all lands (which work has lagged behind) and after determining their productive capacities, potentials and problems, we should decide about different strategies for use and development of different types of soils or land surfaces in different climates and moisture zones in different parts of the country, with a view to their optimum utilisation..

Each type of soil, each piece of land or each tract or watershed (as they now define such locally viable or manageable units) has its own potential or capacity, its own problems, its own particular natural use. Each such unit of land requires certain particular types of inputs or techniques for conservation or development. Therefore, after identifying the various agro-climatic regional, zonal or local land areas, river basins or watersheds and their problems, different suitable or practical steps and strategies need to be adopted for each unit of land.

As far as problem areas like mountainous regions, hill slopes, ravines, waterlogged or flood-prone areas and other lands subject to droughts or the hazards of wind, water or solar erosion or those affected by salinity etc. are concerned, fortunately for us, recent researches, trails and demonstrations have established certain viable technological measures of improvement of control of such areas. Some of these measures need to be applied on a nation-wide basis in private, government as well as common lands.

Unfortunately, at present, almost all States and the so-called progressive or more enterprising cultivators in all regions are trying to compete with leach other in the production of certain 'favourite' or commercially more paying crops regardless of the local conditions or regardless of cost or the wastage of certain inputs. This approach needs modification.

A Multi-grain Buffer Stock

It is possible that the adoption of such an alternative strategy based on proper land use of cultivable lands according to the potential of various land formations and regions may not help us in building up fairly comfortable buffer stocks of rice or wheat. But if the programme, as suggested above, is taken up in right earnest and on a long-term basis, gradually, the vegetation cover would return to almost all the open and barren lands; and there would be adequate production of the most suitable crops on agricultural lands in every separate region or agro-climatic zone.

In that case, we can add to our buffer stock other cereals, including coarse-grains like bajra, jowar, maize, pulses, grams and oilseeds, besides rice and wheat. Perhaps this would also be more in keeping with the old dietary habits and local customs.

It is hoped that once such steps are taken, a stage can be reached when it may not be necessary to depend upon imported foodgrains either to tide over difficult times or the drought years or to build up buffer stocks, as has been done during almost 25 to 27 years since after 1950.

Land Development in Different Landscapes

The wider implications of land development programmes for different areas, river basins, regions, large and small and even mini watersheds, can be appreciated better if we bear in mind the fact briefly mentioned earlier that. by the middle of 1970s, the total net cropped area had increased to about 145 million hectares or nearly 45 per cent of the total geographical area of the country.

This increase came about largely because of the emphasis on grow more food campaigns through extension of the cultivated area and by allotment of government lands or re-allotment of lands retrieved from jagirdars and zamindars with the implementation of land reforms and finally through the *Bhoodan* or *Gramdan* movements. Similarly, the irrigated area also increased. But it cannot be claimed that all the lands under cultivation or irrigation have been duly improved or developed so as to be productive under constant pressure and repeated production of crops.

Regarding the cropped area, it has now been ascertained that nearly 87 million ha. or 60 per cent of it requires some treatment or improvement or careful attention so as to eradicate the problems of erosion or salination or waterlogging and the like. Similar problems but perhaps in somewhat more aggravated form have also been identified in respect of the non-cropped or uncultivated areas comprising of unculturable and barren lands, culturable waste lands, pastures and grazing lands or even forest lands which are being fast depleted or denuded of their greenery and tree cover.

For the country as a whole, it has been estimated (earlier by the ministry of agriculture and later by the department of environment) that more than 50 per cent of the total land surface in the country (or nearly 175 m. ha. area) is in need of proper and scientific management through soil conservation, reclamation, contour bunding and many other requisite measures.

First Priority

Since poor and sick or degraded and neglected lands cannot lead to better production or prosperity, it is imperative that this programme of land development and scientific management must receive our immediate and topmost priority. One may even venture to suggest that if need be, certain other programmes including storage of irrigation water may well be postponed; and the rural works programmes may be reoriented in this direction.

The magnitude of the problem is really immense and calls for early but sustained long-term efforts. Unfortunately, the earlier phase of soil conservation and such allied measures seems to have been overtaken by other schemes or programmes in the name of increased agricultural production through research and extension work or through the development and application of new inputs. But the basic land resource has not been given the attention it deserved. Perhaps, that is why today, with all the efforts in the research institutes or through extension agencies and with all the investments on the production and application of inputs, the country has reached a stage when stepping up agriculture production has become difficult, if not an impossible task. Obviously, we have to revert to the basics and understand that unless we pay due care and attention to the proper and scientific management of our land, it will not be possible for us to break through the vicious circle of stagnation.

There is no country in the world which may have been able to achieve and sustain higher agricultural production levels or modernisation of agriculture without proper land and water management. This may have been through private initiative or through state efforts or through the voluntary or regimented land armies; or whether by mechanical or manual or biological means. Land, as the source of all material wealth, has been looked after to maintain its productive capacity by all the prosperous and food surplus countries.

Land Development and Social Justice

When we talk of distributive justice or social justice in addition to the emphasis on higher production from agriculture or our rural economy; or when we try to achieve these social objectives by introducing land reforms or by distributing available unoccupied government lands or lands declared surplus under the land reforms programmes, we seem to ignore the fact that the persons with limited means, mainly landless agricultural labourers or small and marginal farmers or persons within the category of weaker sections and belonging to the Scheduled Castes and Scheduled Tribes, neither have the means or the wherewithal to make necessary investments on land, nor can they attract sufficient credit, tools or implements and even inputs for looking after their lands, which in most cases, happen to be marginal or sub-marginal as far as productive capacity goes.

And then, because of the agrarian system, the entire land surface, particularly the land under occupational holdings has been continuously broken up and sub-divided into small holdings. This process of fragmentation has been going on because of the growing pressure of population on one hand and the attempt on the part of the well-to-do or even the middle class cultivators to escape the ceiling limits under land reform laws. To these groups have been added the new class of small and marginal farmers to whom lands have been allotted for the first time. (According to 1970-71 Agricultural Census, the average size of a holding in the country was 2.3 ha. though 60.6 per cent of the area under occupational holdings was under holdings of 4.4 ha. or above, and such holdings were about 15.2 per cent of the total). It has to be seen whether and how such small holdings can lend themselves to better management and begin to give higher production. And if so, what must be the conditions or incentives to ensure better management and higher production from such small holdings.

It is often suggested that in an economy based on small and medium sized peasant holdings, the price of agricultural commodities, and some sort of regulation or supervisory control in marketing and facilities for warehousing with the flow of credit etc. will bring about the desired results. This approach has been tried and has also proved successful to some extent; but it has its limited value or usefulness; and perhaps a stage has come when price incentives and such other marketing mechanism or devices may not result in sustained higher production except in the case of the more affluent and resourceful cultivators or in regard to some commercially more productive or profitable crops, which are also subject to seasonal fluctuations.

Land Reforms: Some Essential Steps

We often talk about continuing with land reforms or completing the process which was started in the wake of independence; and in this connection the imposition of land ceilings is placed on top of the agenda though it has been experienced that by now those sections of the farming community which ought to have been brought under the ceiling laws long time ago have managed to find ways to circumvent the effect of such imposition. Besides, not much worthwhile, useful or productive surplus land is likely to be available. In fact, perhaps the real purpose of land reforms (or at least the pending intent of it) can be served if, instead of harping on the imposition of ceilings, two or three other land measures are taken up with imagination and boldness tempered with flexibility according to the changing situation. One such measure should be to eliminate various types of middlemen or gentlemen farmers who do not actually base their living on land (with a few exceptions as in the case of physically or mentally handicapped persons serving men and officers of the armed forces, children and women and so on). It should be enjoined upon all occupational holders or those who have land titles in their name that they must regularly and continuously reside in the villages and as far as possible personally look after, improve and cultivate their land. Another step in the right direction would be to enforce consolidation of holdings which has been given up in many states under the so-called "popular pressure or because of the fear" of the cultivators. It has been noticed that some of the states which have done very well under the green revolution had taken up the consolidation of holdings long time ago and implemented it successfully. It should also be ensured that holdings will not be allowed to be parcelled or segmented any more. Above all, following the relevant provisions in the various revenue acts of different states, it should be enjoined upon all cultivators that they must compulsorily undertake necessary improvement (soil conservation, levelling, drainage etc) works on their land holdings individually or in groups or within the watersheds in which their holdings lie as per advice and recommendations of the concerned technical experts or extension works.

Service Co-operatives

Perhaps, it may also help if serious efforts are made to bring together at least the small and marginal cultivators or those belonging to SC and ST and particularly those who have been allotted lands during the last one or two decades but who have failed to take full or maximum benefits out of this measure into the co-operative fold for arranging and operating their common pool of implements or for arranging small and medium irrigation facilities, for transportation and marketing of their produce as well as for arranging necessary inputs for their mutual benefits. At least this was one of the dreams of our national leaders which has not materialised so far. The next step in this direction can be cooperative or collective farming by pooling of the lands where possible, feasible or necessary.

Dry-land Farming and Watershed Projects

More recently, at the instance of the Union Agriculture Ministry, programmes have been launched in all the states to promote dry-land farming and to cover natural watershed areas under proper management practices. Both these are mutually complementary steps in the right direction.

As per latest reports, a total area of 13.46 million hectares has been covered under dry-land farming practices within the first year of this scheme. Out of this, 12 million hectares have come under "extension approach" so that extension staff can inform and train the cultivators in respect of techniques and practices like use of seed-cum-fertiliser drills, use of drought-resistant and short duration crops, deepening of tanks and ponds, introduction of inter-cropping and double cropping etc.

It may, however, be noted that land improvement and conservation techniques are not yet very prominent in this approach.

However, such work is being attempted, though on a limited scale, in about 3,824 micro watersheds (each covering about 2,000 to 3,000 ha). In these watershed areas, crop production, social forestry, pasture development and horticulture and such other linked activities are being taken up intensively and in a multi-disciplinary manner.

It is also proposed to launch a pilot project costing Rs. 5 crores on propagation of water conservation technology. These are but small and rather belated efforts. We need to cover the entire length and breadth of the country with such programmes.

Watershed Management: First Step Towards Area Development

What is required after the basic survey of the land and water resources and a technical appraisal of their ultimate potential, is the selection and application of the proper technology, which should be economical also, through certain inter-connected and mutually supporting services. Such a programme of development has to be backed by a consistent and long-term public policy to be carried out through the people's representative and technical apparatus of the state.

At a national seminar on watershed management held at New Delhi in January 1980, the demarcation of regions, basins, catchments, sub-catchments and watersheds as suggested by the All India Soil and Land Survey Organisation was accepted.

It was also recommended that, at the time of planning and implementing programme of watershed management "the boundaries of the watershed" should be followed for items relating to the development, conservation, storage and use of water; and for other activities like agriculture, etc. the existing "revenue village boundaries" may be used. These smallest viable units, that is the micro-watersheds, may be taken up for development as separate units or in the form of a manageable cluster project for such units, where such units occur in close proximity to each other.

Comprehensive and Integrated Approach

Whether the programmes which should be comprehensive and complimentary are take up at the "micro level" or at the small or big watershed level, we cannot lose sight of the fact that the smallest unit as well as the intermediate level should really be considered as parts of an "integral whole", which may be a "catchment area" or a "river basin" or an agro-climatic or ecological region. This suggests the strategy of area plans or regional development programmes for development and consideration of natural resources.

The approach will have to be comprehensive and multi-disciplinary, the experts and functionaries of various disciplines or services being placed under a common organisational setting at each different level of planning and programme implementation.

It has now come to be accepted that watershed management is the proper approach towards the conservation and building up, as well as scientific and continued use of the potentials of our land and water resources.

Beneficiaries' Control and Participation

It is also being increasingly realised that for any organisational arrangement for proper watershed management, the active association of the beneficiary participants in an institutionalised and organised manner with sufficient financial, technical and managerial resources at their level is very necessary.

To my mind, the only satisfactory and responsible organisation which may eventually emerge and become responsive to the needs or problems of the local areas can be appropriate forms of a "people's organisation" or "a beneficiary group". We may call it the panchayat or mandal panchayat or the panchayat samiti according to the size of the area and scope of work; but whatever its area or unit, it should have all the technical expertise and financial or managerial resources at its disposal. All lands and local water resources should be left to its management. Only in the case of gross mis-management or negligence, the custody of such common lands may be taken away from the people's organisation (Gram Panchayat, Panchayat Mandal or the Panchayat Samiti as the case may be).

In fact, I would go a step forward further. Not only the common or uncultivated lands, wastelands, ravines and river beds; but also the local grazing and pasture lands (and later on forests as well as the cultivable lands, particularly other than current fallow lands) can also be placed under the care and custody of the local people or their panchayat, mandal panchayat or the panchayat samiti, depending upon the area of a natural physico-geographical or topographical unit of which such lands may constitute a natural watershed.

The management functions should be entrusted to the participants (beneficiaries) or their representatives with the assured support and participation of the technical, professional or financial agencies of the State or other institutions. Unless the people are made responsible, they will not become responsive and vice-versa.

What is being proposed is not something entirely new, because a similar approach was the basis of the original community development and NES set-up of rural development. The same approach has again been repeated with slight modifications in the programme of research and national extension which has been launched with financial support from the World Bank since 1974. The command areas and special areas programmes are also being administrated along these lines, though the beneficiaries are mostly on the receiving end.

Some Pre-conditions for Success

Land management for cultivation or any other purpose on a watershed basis will be successful if some at least of the following conditions are met: (i) If care is taken that the size of area of the land units is large enough to generate the required resources, and where the different types of land use can be planned on an interdependent and multi-disciplinary basis. (ii) The area should be viable, economical and suitable for the application of various technologies. (iii) At the same time, the community of the land-users or beneficiaries, or cattle-breeders or both, or agricultural labourers should also be as closely knit as possible. (iv) The community should be properly motivated and organised. (v) There should be easy access to outside help, guidance, credit and other inputs. (vi) If management leads to some marketable surplus of the agricultural or natural produce, by-products or waste products, proper tie up with the local markets will be necessary. We can add to this list as we go into further operational details.

For larger water-sheds or groups of small catchments and sub-catchments or for small groups of what are called "mini" or "micro" watersheds bigger management units or small project units can be organised as the case may be.

The organisational or the administrative set-up will also have to be such that it can coordinate and supervise all the multi-farious activities relating to agriculture, animal husbandry, soil conservation, water-harvesting and management of inputs, credit facilities, marketing and so on.

At present, we have a slightly different type of field set-up, particularly at the district level, where field units of government departments, co-operative network, and the organisations and development agencies like SFDA and MFAL etc. for small or marginal formers have been set up, largely through government initiative and under government control.

In the long run, the role of the beneficiaries or the people's representatives and the government agencies will have to be completely reversed.

While the guidance, planning initiative and management etc. should come from the people's representatives, it should be the government's function to provide appropriate and economical technologies or technical guidance and wherewithal, with adequate financial support and assistance in subsequent activities like marketing opportunities, along with transferred or delegated power and authority to enforce necessary rules and regulations.

IRDP Strategy and Set-up

Recently, a new development strategy has been initiated under the Integrated Rural Development Programme. Increasing production and productivity in agricultural and allied spheres is one of the key tasks of this programme with which it is proposed to link the centrally-sponsored scheme of National Rural Employment Programme, so that the rural poor may have additional opportunities of employment during the lean season through work-oriented local projects. IRDP also aims at development of resources; and in case of the vulnerable or weaker sections, promotion and improvement of their skills, provisions of credit and marketing facilities and essential minimum needs of the community and the like. In fact, it is a combination or integration of the essential features of a variety of programmes launched in the past and some of which are on-going. Unfortunately, care and development of local natural resources according to their potential are not receiving the attention under the IRDP which they deserve.

Steps have been initiated to establish suitable organisational machinery, but mostly at the district level, to prepare five years development projects for selected blocks in a phased manner. The first starting point of these projects has to be the development of agriculture and allied activities in the primary section, and then take up the potential of development in the secondary or tertiary sectors and to create or bring together the various services, extension, credit, inputs and marketing agencies. A single executive agency is proposed to be set up at the district level. At present, it is likely to be an official agency. But at the same time is proposed to resurrect the old CD or NES block type of organisation and strengthen it with staff for specialised and village level functions. The Government of India has already started giving the necessary subsidy for the IRDP set-ups at the district and block levels.

District Agency: Need for Change in its Role

In view of what has been stressed in the earlier parts of this paper, it is strongly advocated that as soon as possible, the character of the district agency should be changed to make it more representative and responsive or responsible to the needs and urges of the local communities. The district-level specialists of various disciplines should act as their advisors and agents. Ultimately, the "district agency" should be able to hire and fire its own team of experts and technicians rather than remain dependent on the technical and development department or services of the state governments. The same approach should be extended to the block organisation and the next level below, if any.

Local administration on the concept and principles of autonomy is the inevitable direction in which we have to proceed if all the objectives and targets have to be achieved.

Secondly, with regard to the (five years or long period) perspective plan of local area development which is now being attempted at block level, it is strongly urged that the basic concept or approach of comprehensive land and water management on the basis of natural watersheds should be adopted and if need be, block or even district boundaries may be rearranged accordingly.

All the necessary steps which have been enunciated earlier such as survey and inventory of the potential, present-day use and future balanced plan with regard to agricultural, forest, grazing and waste lands, etc. and the identification of special problems of the area must be included in the preparation (as well as the format) of the local plans.

Various Models of Organisations

Lastly, there is need to integrate the various organisational and staffing or servicing models which have been recommended in the context of various development strategies or as per the segmented approach of developing one or the other sphere of activity.

We have the working model of development and technical departments extending their hierarchies to the field level. We have the models being followed under the World Bank supported national extension projects, or the dairy projects. We have the models (rather shadow systems) of SFDA, MFAL, DPAP and such other target population or target area programmes, which are largely based on the core of district administration. We have the models recommended by the Ashoka Mehta Committee on Panchayati Raj and various state or regional committees. Recently, we have worked out the arrangements under the command area and tribal or desert development projects or programmes. Various seminars on watersheds management and the problem of arid and semi-arid zones have also made some recommendations.

Perhaps the common features of these models or recommendations are (a) decentralisation (b) regional area approach (c) support of specialist services (d) people's participation or association of participants with the programmes (e) state level support with funds, technical guidance, credit facilities and so on. These aspects need to be provided for in the future for proposed set-ups for land and water management.

Land Management: Power and Technology to the People

We have now to give primacy to the roles of the local administration and local popular representatives. The approach may be summed up into three working slogans or catchy phrases: (i) Right and power to the people (participants) to manage their own local land and water resources, (ii) Science and technology at the disposal of the local people, and (iii) Man power or technical resources besides adequate funds at the disposal of the local people to enable them to manage their own affairs, plans and physical resources.

Over the years, as various technical services and disciplines connected with different aspects or segmented sub-sectors of our rural economy have developed and gained their own momentum with ear-marked budgets, service cadres and departmental hierachies there has developed an unfortunate tendency to keep each unit of activity and department as separate and distinct from the other as possible. Each field unit has tried to preserve and proclaim its own identity or personality. This has made field-level coordination a difficult and almost an impossible task. Thus, for example, the distance between the soil conservation organisation and field units connected with the supply of credit or agricultural inputs or agriculture extension or minor irrigation has been gradually widening, until no focal point or common meeting ground for common objectives has been left. This tendency has become all the more prominent in respect of some other sub-sectors such as minor irrigation, animal husbandry, cooperatives and the like, which have come under the protective umbrella of special projects.

Revised Organisational Structure: A Proposal

The possible revised organisational structure for the variety of programmes which have to be taken up now on a more integrated and comprehensive scale and as part of interdisciplinary efforts and mutually-supporting activities with the beneficiaries playing a predominant role may be somewhat on the following lines:

(1) For the last two or three decades, the office of development commissioner has been there at the state headquarters, though over the years, his functions have been changed and some of his responsibilities transferred to other secretariat level officers. But he has to be made to function as chief coordinator or the chief leader of all the organisations and agencies connected with agricultural and allied sectors and rural development.

(2) Either under his chairmanship or separately, there should be a coordinating team of heads of technical or development departments, who should function more as technical experts or staff officers rather than as managers or controllers of the budgets,

services, institutions or hierarchies of staff under their control. Apart from their local area functions in association with the local authority, they will remain responsible to the state government.

(3) Similarly, at the regional, district, sub-district and tehsil or panchayat level, the role of the specialists and technical functionaries will have to undergo a big change in respect of the programmes and functions and services which would not go under the full control and jurisdiction of the local authorities.

(4) Experts and technical functionaries may have to (or should) break their links from their present-day parent departments, but may be selected or allotted to work as members of a common team at various levels of functioning, viz., at the state or secretariat level, at head of department level, at the regional and district or sub-district level or tehsil or panchayat level.

(5) At each such level, there should be a team-leader or coordinator of the team and in-charge of the various programmes and projects. He will have to have a well-defined relationship with the appropriate level of supervisory authority or the local popular authority as the case may be.

(6) The role of the parent department may be restricted to cadre management and providing necessary technical support or guidance, enforcement of rules or regulations, or administration of state-run institutions and services, and control of their budget as well as inspection and monitoring etc. of the local area programme. The present-day nearest working model is perhaps the working of the cooperative department vis-a-vis various cooperative institutions.

(7) In a way, break-up of the hegemony of the technical and development departments will help and promote a better and more congenial climate for the emergence of areawise, regional, sub-regional, district or lower field-level units of development manned by multidisciplinary teams of experts and technical personnel.

It is not intended to stipulate the functions or the organisational and staff composition of different levels of operations, except to stress that in the existing set-up, or as per our recent traditions, there is a yet no clear-cut picture of the role of the regional or range-level coordinating and supervisory teams, except perhaps in the recent models of the commissioners for command areas, or the desert development and tribal area development commissioners and the like. Similar steps under other regional development commissioners may be continued, rather strengthened.

Area wise Budgets

Perhaps the biggest innovation which may be introduced should be to provide budget provisions not for the department as such, but for the various regions and other viable and district administration-cum-development units in respect of all the transferred or delegated functions and services. The regional or rural level heads of the local technical agencies should have the right to suggest transfer of funds from one activity to the other depending upon the local requirements. The usual control of the head of the department should be nominal, advisory and of supervisory.

At least in respect of the agricultural and allied sectors, including the programmes connected with the rural development in particular, the budgets should be drawn up on "a local area to local area" or "region to region" basis instead of the present device of providing separate budgets for various departments like agriculture, animal husbandry, forestry and cooperatives etc. and then breaking up the overall department budget for the district, sub-district or the special areas.

18

Problems of Agricultural Marketing in the Desert Areas of Rajasthan

S.P. VISHNOI

In this paper, an attempt has been made to identify problem which are faced by producers in desert areas in respect of marketing of their agricultural or animal husbandry produce. Since the agro-climatic conditions are not suitable for crop farming, inherent constraints inhibit overall productivity in arid and semi-arid areas. As for animal husbandry, desert people and their animals have adapted themselves to these agro-climatic conditions much better and hence animal husbandry is the more important activity. Although a large number of schemes for increasing production and productivity both in agriculture and animal husbandry have been implemented in these areas, yet the economic condition has not materially improved. This seems partly due to insufficient growth of the infrastructure in desert areas and partly due to inadequate attention paid to the marketing of the produce. Various aspects of the process of agricultural marketing have been explored in this paper to identify the problems, likely solutions and to assess the impact of the schemes so far.

Marketing process starts with the decision to produce a saleable commodity. Agricultural marketing involves pre-harvested and post-harvested operations. One would broadly classify these as production, assembling, grading, transportation, storage, marketing and processing. The National Commission on Agriculture has identified four objectives of an efficient agricultural marketing system. These are:

(a) To enable the primary producer to reap the best possible benefits;

(b) To provide facilities for lifting of all produce which the farmers are willing to sell at an incentive price;

(c) To reduce the price spread between the primary producers and ultimate consumer; and

(d) To make available all products of farm origin to consumers at reasonable price without impairing the quality of the produce.

Let us examine, in the context of these objectives, the recent development and the present situation tin arid and semi-arid areas.

II

Since agro-climatic conditions do not suit agricultural operations, creation and development of infrastructure were generally neglected in the past. Poor quality of soil, widespread erosion, insufficient and uncertain rains, lack of irrigation facilities, limited use of improved inputs or new techniques, poor productivity of the labour etc. adversely affect production. Uncertainty further compounds the matter. The agriculturist has become conservative and pessimist; he is reluctant to take any risks.

Unfortunately, not much has been done to improve productivity in these areas. For obvious reasons, these receive comparatively less attention than that devoted to irrigated or command areas. Most of the efforts (and naturally all the breakthroughs) in the agriculture sector are in respect of irrigated crops. Limited research programmes and activities to develop better technology or management practices have not been successful. Whatever has been achieved in various research centres is yet to be successfully brought to the farmers. Producers in irrigated area have access to improved technologies, methodologies and other infrastructural facilities.

Since the use of various inputs in arid and semi-arid districts is much below the state average, desert areas do not interest the commercial agencies and nor do they receive sufficient attention from the government organisations.[1] Thus, the producer starts with an initial disadvantage in respect of production activities.

Natural forces, which adversely affect the productivity and production in agriculture sector in the desert have, however, contributed in their own way to the growth of animal husbandry. Some of the best breeds come from arid and semi-arid areas. Animals have learnt over years to withstand the severity of agro-climatic conditions and yet give satisfactory economic returns to their owners. Vast pastures and even limited water resources have been used by men and animals to their advantage. Unfortunately the current drought conditions reduce the availability of such pastures and water holes. Some producers, under a mistaken notion that crop husbandry is more attractive (either professionally or financially), have tried to divert a part of the land to agriculture, depleting available natural resources for animal husbandry. One of the basic reasons why these animals have been able to withstand the severe agro-climatic conditions is the migratory system. However, certain traditional routes used for transmigration of cattle are no longer available. Various developmental agencies take limited interest in organising and maintaining static and mobile infrastructural facilities along such routes because of the present trend in drought relief operations in which feed, fodder and water is provided for stall feeding. It does give temporary relief but also interferes with the nature's way of culling out the sick, the infirm and the old cattle and of teaching the young ones hardy habits. The thrust of the animal husbandry development activities has been basically to control indiscriminate breeding, to improve the breed by selective/ cross-breeding, better health care, better pasture and feed availability and finally, improvement in marketing system. There has been constant increase in livestock population[2]. However, the attention paid to marketing has been comparatively limited. In respect of production programme, the cattle have been well taken care of in the various dairy projects. In respect of sheep, the emphasis has been on improvement of the quality of wool by selective breeding or cross-breeding. In goats improvement and mutton production quality through cross-breeding or selective breeding has, perhaps, not received sufficient attention. Other animals such as horses, camels etc., in spite of considerable potential, have been neglected by the government agencies.

III

Overall poor economic conditions force producers to devote minimal attention to 'assembly' and 'grading' activities. Since they are in a hurry to market produce, farm level handling itself leaves much to be desired both in respect of agricultural and animal products. Both these products tend to acquire foreign matter at assembly/collection time at the farm level. The loss (both in quality and quantity) is considerable in the absence of adequate care and attention by producer and also due to lack of sustained efforts to develop consciousness in this regard by the developmental agencies. Proper guidance and training could help producers not only to conserve but also to improve the quality. It would certainly enhance the competitiveness of their products.

In respect of 'grading', one finds that apart from commercial crop of cotton, no other agricultural produce is subject to grading in Rajasthan. Even in this respect only 29 per cent of the market arrivals are being graded. The total quantity graded (1,53,873 quintals) is obviously an insignificant part of the total marketable/ marketed surplus.[3] As such any premium which producers may get on account of regular grading is not available to him. However, in irrigated areas (and in respect of major crops), due to extensive purchases by government and private agencies, there is some quality consciousness and as such grading gets automatically introduced. Such organised or large scale purchases of agricultural desert produce *bajra*, lentil etc. are limited or non-existent. There is no effective effort on the part of any agency to introduce grading or quality consciousness in respect of various animal produce. However, animal products (eggs, mutton etc.) produced or consumed in semi-urban or urban areas get almost automatically graded on account of keen competition and because of urban consumers' preference for quality/graded products. Similar products from desert areas not enjoying benefit of quality consciousness of urban markets remain ungraded. Inadequate attention has been paid by development agencies to grading of animal husbandry products except for wool. Total quantity of wool which is subjected to grading before sale is, however, limited and as such a major product (Rajasthan produces 45% of the wool in the country) does not enjoy the benefit which accrue by grading. Grading has been incorporated as an essential feature in respect of milk as all sales and purchases in organised dairy sector are related to the quality of milk. Consciousness in this regard is increasing.[4]

IV

Storage of produce in desert areas is adversely affected as scientific storage of grains and other produce is not being practiced. Due to lower productivity and insufficient production in desert areas, development agencies prefer to concentrate on other areas having larger production. Improper storage at the farm level itself causes considerable losses. At the intermediate storage also, for obvious reasons the agencies responsible for creation and maintenance of storage facilities concentrate on areas with substantially larger production. A brief look at the figure of storage capacities in Rajasthan[5] created either through the national and state agencies or through the cooperative department would show that the bulk of the construction programmes has been allotted to command or irrigated areas. Even in the new programme of rural godowns with the assistance of EEC, desert areas do not get their due share as shown in the following table:

	Number of Godowns Proposed	
	Desert districts	Irrigated districts
50 mt capacity	1018	731
100 mt capacity	395	1987
250 mt capacity	24	97

Source: Registrar, Cooperative Societies, Jaipur.

Adequate storage and warehousing facilities at convenient locations should be available to small producers in arid and semi-arid areas. Selection for construction of godowns must be made by giving certain special weightage to arid areas.

The storage position in respect of animal husbandry products presents an unusual situation. On account of their nature, some of these products require special storage facilities. The storage and intermediate handling before marketing of one of the important cattle products is, to some extent, taken care of by the network of chilling and processing plants established under the dairy development programmes. However, the pace of creation has not been quick and satisfactory nor is the coverage adequate. Regarding sheep and wool, adequate capacities do not exist. Both sheep and goat are handled in the old crude traditional fashion. As for wool, the storage creation has been neglected at the primary stage; only at the secondary or final stage, especially after grading some satisfactory storage facility has been created. No proper attention has been paid to the storage of mutton. As is well known, in foreign countries, the shelf life of mutton is increased by using special cold storage at the production and transshipment points. The lack of such cold storage act as a constraint on the overall marketing activities.

V

The inherent problems of transportation in desert areas plague marketing activities. Scattered population, limited production, scarce resources etc., have meant low economic activity and as such have inhibited the construction and maintenance of trunk and subsidiary roads. This is a vicious circle as this inadequacy by restricting the volume of business in turn aggravates the economic situation. The present network[6] does not adequately serve all arid and semi-arid areas. Even this limited system is susceptible to rapid deterioration due to heavy traffic and also due to severe weather conditions. Rest of the areas is served by gravel or *kutcha* tracks. Vehicles, which are heavy on POL, tyres and other accessories have to be used. In times of general scarcity, overall availability of these inputs for transportation system is itself affected in these areas. All this leads to comparatively higher transportation costs. The transportation even to the primary markets by animal loads, bullocks or camelcarts is time consuming and inefficient and as such costly.

In the absence of assured satisfactory and cheap transportation from the places of production to the main markets (which, in any case, are few and far between)[7] producers are often reluctant to undertake journeys themselves for marketing their limited produce. This has given rise to petty middlemen and traders who induce the producers in inaccessible or difficult areas to part with their produce to them for final sales in regular or periodic markets. Such markets or middlemen have their utility. They do render an important service though at a costly rate. Effective linking of marketing through cooperatives is the solution.

Most of arid and semi-arid areas are not well served by the railways and by the airlines for the same reasons which have inhibited the development of the road transportation. These areas in

Rajasthan are served by meter gauge system which makes transportation slow and hence costly. Even in this network, the movement of produce is tardy due to lack of generation of adequate 'empties' which often depends on overall economic and industrial activity in the area. Movement of cargo by air is not yet popular in India, but in any case these areas have no air service worth the name for passenger or cargo traffic.

Constraints, additional costs, delays etc., identified above adversely affect also the movement of animal husbandry products. Some of these require special type of transportation in the same manner in which they need special storage. In respect of livestock, producers prefer to transport their herds by long and obviously wasteful road journeys. They apparently feel it is a cheap method of transportation but considering loss in weight and also mortality, it is costly. When it comes to products, milk and milk products are, to some extent, taken care of by the arrangements of the dairy network. Slaughter houses have come up close to consuming areas itself as there are no facilities to transport mutton in refrigerated vans etc. Quality and quantity would obviously get affected if such products are moved over large distances, that too, by a slow, inefficient and costly transport system. Wool, on account of its volume is at a disadvantage when it has to move over long distance to processing and consuming plants. Middlemen use these to depress the prices available to producers in primary markets.

Development of transportation facilities in the arid and semi-arid areas has been stagnant for these very reasons for which other development was tardy in the past. Normally, one would have presumed that such a large number of development schemes would have also created additional transportation facilities or improved the old ones to help to move the additional production to places of processing and consumption. Unfortunately, the production programmes and transportation arrangements do not seem to have kept pace with each other. The produce as such cannot move quickly, cheaply or efficiently.

VI

Besides these identified constraints in respect of production, assembly, grading, storage and transportation, actual marketing of the produce has its own difficulties. First and the foremost point is the overall poverty and low economic activity in arid and semi-arid areas which is responsible for weak and unsatisfactory demand for different produce and goods (both local and obtained from outside). Big cities, as high consumption areas, with reasonably high and stable demand stabilise prices and thus help the growth of markets. For historical and geographical reasons, such markets are limited in arid and semi-arid areas and new ones are not coming up and thus these act to the detriment of producers. Rajasthan has additional disadvantage which perpetuated as a result of the partition. Certain traditional markets were lost in the partition and long flourishing routes got disrupted. Producers, due to their inherent weaknesses, could not capture new markets in the other areas.

The total agricultural production is these areas is normally insignificant. Bulk of this is either consumed or stored away for difficult times resulting in limited marketable surplus and reduced marketing activity. Producers have also not been able to build up a lobby for their produce at the national or state level. Advantages and benefits available to other producers (of major agricultural produce such as wheat, paddy, cotton, sugarcane, jute etc.) are not available to them. There are no important traders, processors of consumers to espouse their cause. Some even attribute the overall fall in the production of such agriculture produce (say coarse grains) to these factors.

Animal products are mostly primary raw materials and as such their markets are effectively controlled by middlemen, producers of finished goods or quite often by consumers to the detriment of the interest of primary producers. Processing units and consuming areas are away from producing areas. Main markets have developed mostly in and around these processing or consuming areas. Thus, the price structure and the marketing of surplus milk are dependent on the need of consumers of metropolitan cities like Delhi. Wool sales are basically in and around the carpet producing centres or in and around old traditional markets like Beawar. Carpet manufacture and middlemen dominate these markets. Major markets for sheep, goat and mutton are Bombay, Ahmedabad and Delhi. Similarly, markets for hides and bones are either in Delhi or in Bombay.

The traditional indebtedness of producers also plays a constraint on the marketing system. Though the holding may be large, small producer produces only limited quantity which he must quickly unload in even unfavourable situation in order to meet his debt obligation. Middlemen normally prefer to move the produce for better price in the markets or consuming area outside adversely affecting the local markets. The cooperative institutions are there to support such producers by linking of credit with marketing but unfortunately the progress of the movement in these areas is still not satisfactory. Coverage itself is low, and adverse agro-climatic conditions affect activities and viability of cooperative societies. Total turnover (in respect of marketing of produce) by cooperative marketing societies in the desert districts is considerably lower than in the irrigated districts.[8] Benefits, which should follow from an effective linkage of credit with marketing, have not thus accrued. As for animal products, the cooperatives in the dairy field handle the bulk of the marketable surplus. Average handling is around one lakh tonne out of an estimated marketable surplus of six lakh tonnes and total production of 31.1 lakh tonnes (figures per annum).[9] The position of wool sector has not improved; total annual handling by the cooperatives is around 5,000 kg.,[10] out of a total production of 13 million kg. Similar situation prevails in cooperative marketing of livestock, mutton, eggs etc. There is need to adopt an aggressive strategy to help producers.

Banking facilities (both for regular and for agriculture and allied sectors) which have expanded extensively and rapidly elsewhere are still not satisfactory in arid and semi-arid areas. Credit availability and use are limited.[11] Another constraint in these credit facilities is of course the insufficient warehousing structure. Inspite of various credit assistance programmes, producers mostly do not enjoy the benefit of cheap and easy credit.

VII

On account of identified poor quality of produce, insufficient production, difficulties in storage and transportation of produce, institutional purchases are small as government and private procurement agencies are reluctant to operate in arid and semi-arid areas. Lack of powerful lobby at the national or any other level once again acts against the interests of these producers. In other areas, the various procurement agencies (FCI, CCI, etc.) set up a large number of purchase and procurement centres and even offer facilities and monetary incentives to mop up surplus. Perhaps these agencies look at these centres in desert areas from a narrow commercial angle losing sight of any likely advantages to producers. Insufficiency or lack of these operations by various agencies leads to instability/fall in the open market prices and the benefit of announced prices does not usually accrue to these producers.

Some of these aspects should have been taken care of by the regulation of agriculture markets. Unfortunately, the main thrust of regulation activities has been in urban or semi-urban areas or in areas where the production is substantial. The market committees in arid and semi-arid areas have limited turnover and as such have limited income. Primary and main markets have not developed sufficiently. Under the present system, turnover and financial viability criteria are uniformly used for declaring market yards and sub-yards. Many committees in desert areas lose income as large scale sales take place in unregulated small and periodic markets which do not qualify for notification under these criteria. Such committees are unable to develop their own infrastructure or contribute substantially to the overall development in their hinterland. For example, the market committee of Ganganagar has been able to complete 54 kms. of rural roads besides other facilities and has got another 24 kms. of such roads in hand. As against this, most of the marketing committees in the desert districts have not even been able to taken up the construction of rural roads and those who did, have not been able complete any. Same is true in respect of other facilities which such committees can and should provide to producers.[12] There is a Master Plan for rural roads. Its execution contemplates substantial contribution by local committees also. Obviously, roads identified in desert areas may never be taken up as the concerned market committee may not have the requisite funds. In areas where the dairy programme has succeeded, milk routes to facilitate collection and marketing of milk, particularly in desert areas have, however, been developed. The state agricultural marketing boards were to create a "Development Fund" and use the same for the benefit of such weak marketing committees. Such a fund with a corpus of Rs. 50 lakhs is reported to have been created by the Rajasthan State Agricultural Marketing Board but it has not been used to assist the weak marketing committees in desert areas.

The regulation of markets, which was progressing satisfactorily in the state, received a setback when 66 commodities were deregulated on 23 April 1977. Most of these were the main produce (both agriculture and animal husbandry) of desert areas. Their deregulation is bound to have a long term adverse affect. Both producers and market committees would lose. Although it may be an isolated instance yet it illustrates a comparative lack of appreciation of the problems and the inherent weakness of arid and semi-arid areas.

VIII

In respect of construction of new market yards through institutional finance, lesser attention appears to have been paid to the desert area markets. Published statistics show that in Rajasthan, most of the schemes and bulk of funds have been allotted to the irrigated or command districts where in any case there were already adequate facilities in existence. The main reason for this 'discrimination' may be that in those areas because of higher assured production, better returns on investment were anticipated and as such the schemes were more acceptable to the financing institutions. Unless more positive criteria favouring desert areas are adopted for such investment the discrimination may persist.

IX

The picture is slightly different in respect of marketing operations of milk. Milk producers have suffered exploitation by petty traders and local merchants. A better and assured marketing of the surplus milk has now been organised by the creation of extensive infrastructure for collecting raw milk from producers, thereby eliminating middlemen. The organised cooperative dairy sector now handles nearly 20 per cent of the marketable surplus.

The various animal husbandry development programmes have repeatedly laid emphasis on the importance which must be attached to cattle fairs. Since times immemorial these have been the marketing outlets for different livestocks of arid and semi-arid areas. However, though these fairs bring substantial revenue to the state government and to the different local bodies, yet not much has been done to upgrade and improve these fairs and to make them more beneficial to producer. On the basis of a suggestion made in the perspective plan of Rajasthan drawn up by the NCAER, a Master Plan for improving all such fairs in a phased manner is reported to be under preparation. It would go a long way in helping producers.

X

Another aspect in respect of agricultural marketing which needs attention, pertains to processing. For historical reasons and other identified reasons, processing facilities for the produce have mostly come up not in the producing areas or in their vicinity but in close proximity to the consuming areas. In respect of processing facilities at producing or consuming end, one has to take into consideration the relative benefits and costs. Setting up these in producing areas leads to reduction in transportation cost due to reduced weight and volume of finished products. Another aspects is the availability of by-products in the same areas. Overall industrial and economic activities means more employment avenues. On the other hand, such raw material producing areas are normally deficient in infrastructure. The development of such infrastructure requires considerable investment. This may also mean under-utilisation of infrastructure already developed elsewhere. Planners may like to have a happy mix which does not discriminate against arid and semi-arid areas having regard to their identified constraints. However, most of the Gwar-gum factories, tanneries and leather factories, carpet factories, 'dal' and oil mills etc., are situated at far away places and even outside the state. These far-off locations exert depressing effect on the prices in the producing areas and as such adversely affect the producers. Not only private entrepreneurs but also by some of the government and semi-government agencies seem to have discriminated by preferring areas close to consumption points.

While producers in the vicinity of urban areas or high consumption points are able to dispose of their production quickly, these producers suffer from glut specially in respect of perishable commodities. Very limited work has been done so far to exploit the potential of preserving or processing (deep-freesing, dehydrating, cooking and canning etc.) the perishable commodities with a view to extending their shelf life. Only in respect of milk, it is being done by extracting fat and drying skim milk into powder to extend the shelf life. Development of such processes and their extensive adoption in arid and semi-arid areas have not received the attention these deserved from Government agencies. Commercial ventures have preferred to concentrate on fruits and vegetables and livestock items in the hills or in industrially advanced areas (J&K, U.P., Karnataka etc.).

XI

Market information, market data and market research are very important to any producer. Unfortunately in this regard also, the producers in the desert area are at a disadvantage. They operate mostly in primary markets which are not fully developed. The turnover is limited, their market committees are weak and there is hardly any commercial organisation or lobby helping them. Since the producers do not get the benefit of timely information or analysis, middlemen or traders try to manipulate dearth of the same to their own advantage. Akashwani which carries a bland market prices announcement, could help by having regular analytical talks which may educate producers and help them to get a better price for their produce.

XII

Primary producer is ordinary subject to many disabilities and desert area producer suffers from additional constraints. He does not reap possible benefits of his produce as the entire system seems to be loaded against him; he produces a small quantity of indifferent quality due to agro-climatic conditions and also due to his poor management in an insufficient and inefficient infrastructure. Overall indebtedness, inadequate credit arrangements, lack of suitable storage ware housing and grading facilities force him to part with his produce at a low price at an early date, maybe to a middleman. The lack of developed or regulated market facilities, insufficient transport facilities, locational disadvantage of processing units and general unawareness of market data, its forces and their operations effect his competitiveness and reduce the financial benefits which should rightfully flow to him due to increase in production or productivity.

XIII

Inspite of the attention paid by the different agencies to the desert areas, the results have not been spectacular. Perhaps the total funds allocated for the development of the arid and semi-arid areas are insufficient. These funds are a sub-allocation out of the funds meant for agriculture and allied sectors. Inspite of a greater degree of poverty, inequality, unemployment and lack of opportunities in these areas, the allocation and sub-allocation are not of adequate level. Even in the new Five-Year Plan, the total proposed allocation for the specified arid and semi-arid areas is small. No specific thrust for the creation and maintenance of additional infrastructure has been identified in respect of agriculture and allied sectors.

Another reason seems to be the approach adopted in the past for tackling the problems of these areas. Normally it is claimed that the 'sub-sector' approach for planning should not be adopted in respect of a sector or of a project where the infrastructure is weak or is likely to be a constraint. It has been seen that in arid and semi-arid areas, the development of infrastructure in the past has been neglected and even in the recent past, its development has not been satisfactory. Maybe, this is the reason why inspite of specific schemes for increasing productivity and production in different sectors, the development has not been satisfactory and the desired benefits have not accrued to the producers or the area. In respect of certain items and programmes, even of the 'area', a 'functional' approach has been adopted (SFDA, DPAP, DDA etc.). However, in these all the schemes were small or the areas in which the schemes were attempted were small. The infrastructural problems were spread over vast areas. The schemes and the programmes themselves were unable to overwhelm the infrastructural constraints. These were also not otherwise identified and as such adversely affected the final results. What was required was, perhaps, a regional-cum-subsector oriented programme.

XIV

The problems of desert present different situations both at the planning and at the execution level. Magnitude and ramifications are many and these may tend to dishearten the planners and the administrators alike. However, the experience of the successful dairy programme has shown that where the approach has been coupled with strengthening of infrastructure through a widespread regional area approach the results have been more satisfactory. Integration of various development programmes and schemes in this manner with more specific attention to the infrastructural constaints seem to be the solution.

NOTES

1. *Fertiliser and Allied Agricultural Statistics of North Zone* (1979-80). New Delhi: Fertiliser Association of India, p. 13.

2. Government of Rajasthan, Directorate of Economics & Statistics, *Basic Statistics*: Rajasthan (1979), p. 71.

3. Rajasthan State Agricultural Marketing Board, *Krishi Vipnan Pragati Path Par*, 1980, pp. 40-41.

4. Annual Progress Report of Rajasthan Cooperative Dairy Federation Ltd. (1979-80).

5. Annual Progress Reports of Rajasthan State Warehousing Corporation, Rajasthan State Agricultural Marketing Board, Cooperative Department and Food Department (1979-80).

6. Government of Rajasthan, Directorate of Economics & Statistics, *Basic Statistics*: Rajasthan (1979), pp. 201-2.

7. Rajasthan State Agricultural Marketing Board, *Krishi Vipnan Pragati Path Par*, op. cit.

8. Government of Rajasthan, Cooperative Department, *Cooperative Movement in Rajasthan: Important Statistics*, 8974-197, 1980, pp. 62-66.

9. Government of Rajasthan, Planning Department, *Draft Sixth Five Year Plan 1980-85*, 1980, p. 3.

10. Directorate of Sheep & Wool, Government of Rajasthan, Jaipur.

11. Government of Rajasthan, Planning Department, *Directory of Bank Offices in Rajasthan.*

12. *Krishi Vipnan Pragati Path Par,* op. cit., pp. 268-304.

13. *Krishi Vipnan Pragati Path Par,* op. cit., 1980, pp. 34-35.

14. NCAER, *Perspective Plan of Rajasthan*, (1974-1981).

15. Government of Rajasthan, Planning Department, *Draft Sixth Five Year Plan: 1980-85* and *Annual Plan: Rajasthan* 1981-82.

19

Optimisation of Crop Production in Desert Land

A.N. LAHIRI

Optimisation of plant production in arid and semi-arid areas essentiallly rests on the effective implementation of the principles of conservation. Hazards of wind erosion ask for appropriate soil conservation measures, low and erratic precipitation warrants moisture conservation in sandy soils which are abundant in this area, poor soil fertility conditions require methods for its maintenance and improvement, meagre irrigation waters need effective management conditions for optimum crop production with minimum of water, and plants which often show limited growth due to environmental constraints, deserve a rational utilisation and protection from biotic interferences more than anywhere else. In this regard, there is also a general realisation that only the plant component and the soil component of the overall soil-water-plant-environment complex are liable to some extent to human manipulations and thus technologies for improvement of plant production, particularly in rainfed areas, should largely stem from these two specific areas.

In view of these, an attempt has been made to reflect the results of researches in these directions undertaken here.

Problem of Wind Erosion

Pastoral or silvi-pastoral land use is generally deemed to be the best for reducing the hazards of wind erosion in arid areas. Table 1 indicates that in Beechwal area of Bikaner, large losses of soil and nutrients occurred within 75 days (April-June 1977) of summer due to wind erosion where the soil was left bare after cropping. Erosional losses were markedly reduced where bajra stubbles were left while there was no loss, but a soil deposition in the grass land. The data also reveal that large movement of sand from bare dunes could be completely restricted by their stabilisation through introduction of trees and grasses.

Table 1

Nutrients Depletion/Addition as Affected by Sand Movement under Different Land Use Conditions

Treatment	Sand depletion/ addition t/ha	Organic matter	Total N	Mineral-N (NO_3+NH_4)	Total P	Total K
Beechwal						
1. Bare sandy plain	−1449.0	−724.5	−115.9	−44.9	−231.8	−2159.0
2. Bajra stubble	−22.5	−38.3	−2.25	−0.7	−3.3	−34.9
3. Grass	+15.5	+33.8	+1.62	+0.4	+2.7	+22.4
Udairamsar						
1. Bare unstabilised sand dune	−5560.5	−2780.0	−333.6	−166.8	−1000.8	−6282.8
2. Stabilised sand dune	+155.5	+2121.0	+227.3	+30.3	+333.3	+1818.0

− indicates depletion

+ deposition

Soil-Plant Interaction in Silvi-pastoral System

If the merit of silvi-pastoral land use is accepted, it is necessary to ascertain which of the adapted tree species will be most suited for such programme particularly in the context of their relative influence on the soil and growth of grasses.

In view of the importance of this issue in the land productivity of arid regions, a study was undertaken on the relative effects of different tree (*P. cineraria, T. undulate, P. juliflora, A. lebbek,*

A. Senegal) on the soil moisture, population of soil micro-organisms, soil fertility build up and the growth of the ground cover for ascertaining the suitability of these trees in silvi-pastoral systems.

It has generally been found that the moisture regime remained higher under *P. cineraria* and *T. undulata* than under rest of the trees. This phenomenon caused either due to differential stem flow or root growth behaviour profoundly influenced the population of soil micro-organisms which contribute towards the building up of soil fertility. Table 2 indicates that the soil beneath *P. cineraria* harboured the maximum population of bacteria, fungi and actionomycetes, as compared to other tree species, and the population of nitrifying bacteria was also fairly high. Populatin of micro-organism, as such, below *P. juliflora* was the least. The build-up of soil fertility could be correlated with the microbial population of the soil. Table 3 indicates that the fertility build up in terms of increase in organic carbon and nitrogen decreasing order by *T. undulate, A. lebbek, A. Senegal and P. juliflora.*

Table 2

Population of soil micro-organism under different trees

Sp.	Bacteria $X10^5$	Fungi $X10^3$	Actino-mycetes	Nitrifying bacteria (MPN)
P. cineraria	32	29	16	1,100
A. lebbek	22	18	11	1,300
T. undulate	25	20	12	900
P. juliflora	20	16	10	700

As a result of the foregoing soil-plant interaction, the dry herbage yields under *P. cineraria, T. undulate, A. lebbek, A. Senegal* and *P. juliflora* were of the order of 2.3. 1.66, 1.32, 0.78 and 0.85 t/ha respectively. It has further been reported that the leaf litter which accumulate beneath *P. juliflora* contains inhibitors which not only restrict germination but also the root and shoot growth of germinated seedlings of a number of species.

All the foregoing facts indicate that multi-disciplinary researches may provide a fairly reliable information on the relative benefits of different tree systems.

Table 3

The Status of Organic Carbon and Nitrogen in the Loamy Sand Soil Under the Tree and Grass Plantation during 1975

Observa-tions	Bare	Site	Tree plantations								Grass plantations					
			P. cineraria		*T. undulata*		*A.lebbek*		*A. senegal*		*P. juliflora*		*L. sindicus*		*C. ciliaris*	
	0-15 cm	15-30 cm	0-15 cm	15-30 cm	0-15 cm	15-30 cm	0-15 cm	15-30 cm	0-15 cm	15-30 cm	0-15 cm	15-30 cm	0-15 cm	15-30 cm	0-15 cm	15-30 cm
% O.C.	0.14	0.12	0.33	0.17	0.28	0.19	0.27	0.13	0.27	0.16	0.22	0.18	0.20	0.17	0.18	0.14
			(136)	(42)	(100)	(58)	(93)	(8)	(93)	(33)	(57)	(50)	(43)	(42)	(7)	(16)
% N	0.02	0.02	0.042	0.026	0.041	0.024	0.036	0.17	0.039	0.024	0.032	0.028	0.016	0.02	0.024	0.02
			(110)	(30)	(105)	(20)	(80)	—	(95)	(20)	(60)	(40)	—	—	(20)	—
C/N ratio	7.0	6.0	8.0	7.0	9.0	8.0	8.0	7.0	7.0	7.0	6.0	6.0	12.0	8.0	7.0	7.1

Note: Figures in parentheses indicate increase over control (i.e. bare site).

Table 4

Soil Moisture Status of Unstabilised and Stabilised Dunes at Different Times of the Year 1963-64

Area : Barmer

Depth (cm)	**March** **% moisture**		**September** **% moisture**		**June** **% moisture**		**January** **% moisture**	
	Unstabi-lised	Stabi-lised	Unstabi-lizd	Stabi-lised	Unstabi-lised	Stabi-lised	Unstabi-lised	Stabi-lised
0-5	0.3	0.1	0.00	0.1	0.4	0.8	0.90	0.12
5-15	0.9	0.3	1.95	0.8	0.5	0.8	0.50	0.32
15-30	3.2	0.8	2.32	0.3	1.5	0.8	1.53	0.16
30-45	2.8	0.4	2.85	0.7	1.7	0.9	2.45	0.43
45-60	2.6	0.8	3.12	0.8	2.1	0.8	2.13	0.38
60-75	3.5	0.5	2.65	0.6	2.7	1.4	2.38	0.33
75-90	2.8	0.7	3.33	0.9	3.9	1.7	2.17	0.61
90-105	2.6	1.1	4.33	0.9	2.2	1.3	3.42	0.40
105-120	3.3	1.6	3.85	1.1	4.3	0.96	4.28	0.77
120-135	3.2	1.7	4.10	1.3	3.3	1.02	4.76	0.66
135-150	3.6	1.5	4.22	1.6	3.4	2.6	3.76	0.91
150-165	4.1	2.0	5.11	1.5	3.2	3.1	4.22	0.22
165-180	4.7	1.8	5.85	1.4	3.5	2.0	5.50	1.30
180-198	5.1	2.1	5.00	1.6	4.1	2.2	4.20	2.20
195-210	4.8	1.9	5.10	1.7	5.0	1.9	5.00	2.30
Moisture content in mm upto 105 cm depth	40.96	10.22	44.75	10.3	33.25	16.89	33.52	6.58
Moisture content in mm upto 210 cm depth	105.55	38.59	119.53	33.27	93.58	47.93	104.9	27.66

Consideration of Soil Moisture Situations and Plant Water Use

Among other factors, limitation of soil moisture is by far the most serious constraint for plant production in dry regions. However, soil moisture situation may vary widely from one region to the other due to a number of reasons, such as, variation in precipitation, topography, slope, soil conditions, canopy character and rainfall interception by the existing vegetation and the prevailing pattern of evapo-transpiration.

The dune systems, in this area particularly, provide an excellent example of the impact of vegetation on the soil moisture status. Experimental data indicated that most of the desertic vegetation (except a few trees like *Salvadora oleoides*) transpire at high rates which often exceed 1200-1500 mg/gm/hr. As a consequence the soil moisture of vegetated stabilised dunes mostly remain low. But in contrast, soil moisture contents of unstablised dunes remain high principally due to the absence of moisture loss through plant transpiration. Mulching action of dry surface sand contribute also in conservation of moisture in the profile. The extent of difference between the stabilised and unstabilised dune have been indicated in the Table 4. This conserved moisture of unstabilised dunes obviously contribute much towards establishment of vegetation for their stabilisation.

While considering silvi-pastoral system, it should be reckoned that only the young trees with limited root system will compete with grasses for soil moisture. But, established trees of this area mostly have deep root system and thus hardly ever compete with grasses for soil moisture. Studies on *P. cineraria* community have indicated that the tree component (density of 50 per ha) used about 222 mm/ha annually from deep soil layers while the ground cover of grasses consumed about 162.9 mm during their growing period (July-September) mostly from upper soil layers. Therefore, in established silvi-pastoral system there is hardly any competition between the grass and tree components.

This situation may not, however, be true in all cases and will depend much on the root growth behaviour of the tree species and soil conditions. For instance roots of *Acacia tortilis*, unlike *P. cineraria*, do not penetrate the *kankar* pan in the course of growth and thus in shallow soils may often trail along in the upper layers of the soil up to a long distance. It is obvious that the tree will compete with grasses for moisture as well as for nutrients under such situations.

Management Conditions for Grasses

Considerable information has already been generated at the CAZRI regarding the management conditions of different categories of grass lands, such as, protection, soil conservation measures, reseeding, rations utilisation etc. In this regard, it has been found that the management conditions may be tailored to meet the demands of high forage production or high protein yield per unit of area.

Larger interval of cutting/grazing (30 to 60 days' interval) with low doses of nitrogen (20 kg N/ha) increased the forage yield of desert grasses like *Lasiurus sindicus, Cenchrus ciliaris* and *C. Setigerus*. But for higher leaf protein yield shorter cutting/grazing interval (10 to 20 days) is necessary (Table 5).

Table 5

Effect of Cutting/grazing Interval on the Forage and Leaf Protein Yield of Range Grasses

Cutting interval (days)	*C. setigerus*		*C. ciliaris*		*L. sindicus*	
	Dry matter (q/ha)	Leaf protein (kg/ha)	Dry matter (q/ha)	Leaf protein (kg/ha)	Dry matter (q/ha)	Leaf protein (kg/ha)
10	5.96	62.53	5.78	80.19	5.51	40.86
20	8.53	60.38	8.90	67.99	6.31	68.96
30	10.49	58.42	12.80	82.85	9.16	45.15
60	12.18	45.34	19.02	70.90	13.16	45.15

It has further been found that *L. sindicus* particularly has the capacity for displaying substantial growth during the off-season (November to May) despite the low soil moisture condition. It seems again that management conditions adopted during the summer growth profoundly influences off-season growth of *L. sindicus.* In a good rainfall year, the off-season forage yield from grass may often be over a tonne per hectare.

It may be mentioned in this regard that generally application of nitrogen, longer cutting interval and larger cutting height which contribute often towards larger forage yield also increases the consumptive use of moisture as well as the water use efficiency of these grasses (Table 6).

Table 6

Effect of Nitrogen, Cutting Interval and Height of the Water Use (mm) and Water Use Efficiency (kg dry matter/ha/mm water used) in C. setigerus, C. ciliaris and L. sindicus during 1976 (precipitation during experiment=537.7 mm; total precipitation=639.7 mm)

Treatments	*C. setigerus*		*C. ciliaris*		*L. sindicus*	
	Consumptive use (mm)	Water use efficiency (kg/ha/mm)	Consumptive use (mm)	Water use efficiency (kg/ha/mm)	Consumptive use (mm)	Water use efficiency (kg/ha/mm)
(a) Nitrogen doses						
0 kg/ha	151.8	10.40	173.0	13.73	180.0	11.36
20 kg/ha	176.3	12.96	182.5	13.98	190.6	12.33
40 kg/ha	191.3	11.85	176.6	18.34	194.4	15.71
(b) Interval of cutting						
10 days	175.7	11.80	167.9	11.74	183.6	5.89
20 days	174.0	9.86	191.8	14.15	173.6	10.35
30 days	177.3	13.97	171.7	17.42	200.1	12.53
60 days	182.5	12.00	182.2	18.11	206.2	18.28
(c) Cutting height						
5 cm	158.5	12.81	171.8	17.03	183.6	12.13
10 cm	185.5	11.07	178.1	14.86	189.1	11.58
15 cm	175.5	11.85	186.7	14.14	192.3	11.58

Thus, moisture conservation measures in grasslands of arid areas assumes great importance. Studies undertaken in the rangelands of this Institute located in Jaisalmer (178.5 mm average rainfall), Bhopalgarh (316 mm), Palasana (468.6 mm), and Jadan (411.0 mm) indicated that contour bunding (75 cm height and 80 cm vertical spacing) combined with contour furrowing (10-15 cm height or depth and 100-125 mm vertical spacing) generally improves the forage yield associated with increase in soil moisture status as compared to the control or only bunding or furrowing (Table 7). There were, however, variation due to soil, rainfall and range conditions.

Table 7

Percentage Increase in Soil Moisture Over the Control (based on peak observed values) in Different Moisture Conservation Treatments and the Average Forage Yield from the Grassland in Jaisalmer during 1966

Treatments	% increase in soil moisture over the control at indicated depths in cm			Forage yield (q/ha)
	0-5	5-15	15-25	
Contour furrow	100.4	75.6	40.0	3.04
Contour bund	45.6	18.9	23.6	2.75
Contour furrow and contour bund	129.5	90.6	47.2	4.06
Control	—	—	—	1.44

Conservation Farming

In view of the high density of human population in the Indian desert, the practice of cropping cannot, however, be entirely eliminated. Thus, techniques of conservation farming have been developed. In wind strip cropping method, 30 metres of crop strips are alternated with 5 metres of grass strips. Both are kept at right-angles of the prevailing wind direction. Table 8 indicates that protected crops generally gave higher yields.

Table 8

Grain Yield from Protected and Unprotected Strips (average of 8 years)

Crop	Yield in kg/ha	
	Unprotected	Protected
Bajra	715.5	758.2
Moong	273.5	319.0
Moth	391.1	393.1

Table 9 again shows that where stubbles were left, soil loss was reduced and the yield of subsequent bajra crop was improved. It seems that the stubble heights of 30 cm gave the highest yield but the soil loss, due to wind erosion, was minimum where 45 cm stubble or the whole stalk was left.

Table 9

Bajra Yield and Soil Loss Under Different Residue Management (average of five years)

Stubble height cm	Soil loss q/ha	Grain yield kg/ha
No stubble	75.86	519.4
15	55.41	526.0
30	57.88	576.7
45	24.75	536.9
Whole stalk	26.31	545.7

Drought Avoidance and Conservation of Soil Moisture

Agricultural production in rainfed arid and semi-arid areas is highly unstable mostly because of frequent droughts. Four principal types of drought have been commonly encountered in nature. In one case (as found in 1968), early rains followed by prolonged dry period, progressively decreased the soil moisture reserve and thus the crop growth was adversely affected. In another situation as in 1969, rains failed completely and thus soil moisture remained extremely low throughout the growing period of the crop. This condition frequently occur in the desertic regions. In 1970, plants experienced moisture stress at the critical period of the reproductive phase while in 1971, plants were subjected to sporadic droughts due to erratic precipitation. The latter conditions of drought are often encountered in areas which otherwise may receive high rainfall. Studies on the effects of drought on plant processes, such as, growth, yield, nutrient uptake, metabolism, enzyme activity etc., have led to the conclusion that adequate soil fertility conducive for optimum vigour of plants helps to obtain substantial yields despite soil moisture stress under sporadic drought conditions.

The undermentioned technologies have been developed for soil moisture conservation and for the avoidance of these kinds of drought.

(a) Sub-surface Barrier

Placement of sub-surface barrier (ca. 2mm thick at 60 cm depth) in sandy soil conserves soil moisture in the root zone and thus brings about a higher mineralisation of nitrogen, increases nitrogen uptake by plants and improves the crop yield (Table 10).

Table 10

Effects of Asphalt Barrier on Water Use, Yield, Mineral-N of Soil and N-composition of Bajra during at the Central Research Farm of CAZRI

Treatment	Mineral-N (ppm) (0-45cm)	% N in grains	% N in straw	N-uptake (kg/ha)	Grain yield (kh/ha)	Water use (mm)	Water use efficiency (kg/mm/ha)
Without barrier	33.6	1.51	0.43	29.08	11	188	5.85
With barrier	56.0	2.00	0.59	53.47	17	284	6.0

Table 11

Effect of the Addition of Pond Sediment on N Utilisation, Water Use, Growth and Yield of Greengram

Pond sediments added (tonnes/ha)	Grain yield (q/ha)	Water use (mm)	Water use efficiency (kg/mm/ha)	Total N (%)	Mineral N (ppm)	Increase in N	N in grain	N uptake in grain (kg/ha)
0	15.7	200	7.85	0.031	7.9	10.8	3.23	50.7
38	19.7	230	8.56	0.039	11.2	15.0	3.87	75.2
76	21.5	248	8.66	0.145	13.4	16.0	3.88	83.4
114	18.7	252	7.72	0.050	13.5	13.8	4.14	77.4
152	18.4	332	5.54	0.053	14.5	6.0	4.16	76.5
LSD 0.05	29.1							

(b) Use of Soil Amendments

Moisture retention of sandy soils could be also improved by mixing of pond silt (obtained from desilting of village pond) to the surface soil. This had favourable influence on nitrogen utilisation, water use, growth and yield of crops (Table 11).

(c) Use of Mulches

Use of polyethylene and bajra husk mulches also contribute towards conservation of soil moisture and increases the grain and straw yield of bajra with an associated improvement of water use efficiency (Table 12.)

Table 12

Effects of Mulches on Water Use Efficiency, Dry Matter Production and Grain Yield of Bajra (BJ 104)

Treatment	Grain yield (q/ha)	Straw yield	Water use (mm)	Water use efficiency (kg/mm/ha)
Polyethylene	29.0	35.0	279	10.4
Bajra husk	23.0	30.0	269	8.5
Control	17.4	26.0	291	6.0
LSD at 5%	1.0			

(d) Runoff Farming

Runoff water from artificial catchments also help in stabilising the yield of crops grown in the adjoining cultivated field on sandy flat land. Different ratio of catchment to cropped area was tried and 0.5 ratio was found to be the best (Table 13).

Table 13

Yields (kg/ha) of Different Crops in Bad Years on Flat Surface (control) and Under Runoff Farming (0.5 catchment to cultivated area)

Year		Rainfall (mm)	Bajra	Sorghum	Moong	Guar	Cow-peas	Til
1972	Control	167	0	0	229	255	171	147
	Runoff		395	253	258	234	261	249
1974	Control	117	0	0	0	0	0	0
	Runoff		96	0	121	102	0	36

Optimum Use of Water in Irrigated Crops

The highest possible evapotraspiration of a crop completely covering the soil (the potential evapotranspiration) is determined by the evaporating capacity of the atmosphere, as long as soil is moist. With decreasing soil moisture contents, evapotranspiration becomes dependent on the available moisture. The plant plays only a passive role in the process of transpiration, serving as a means of transport for water from soil to air and adapting itself to the circumstances by closing or opening the stomata.

Although to a certain degree plant production is related to the total amount of water used, the increase in production may no longer be dependent on an increase of available water after a certain level of production. More water at that point may not contribute to the yield because of certain constraints of inputs, such as, soil fertility, selection of variety, etc. Often, however, amount of irrigation water required is determined only on the basis of evaporating capacity of the atmosphere. Since there is a prospect of achieving different levels of production at the same level of water use by optimisation of the inputs, the study of input manipulation assumes a singular importance in arid areas where water for irrigation is limited. It has been found that substantial and economic yields can be achieved in this area by optimising the inputs like nitrogen and phosphorus at a low level of water availability.

The foregoing adjustment of irrigation to production would in several cases lead to the possibility of extension of the irrigated area. Although under such circumstances crop yield per unit area may decrease, but the total production from a project area may increase. The data provided in Table 14 indicate that a marked increase in total productivity of different crops may be possible by using a given amount of water over a larger area.

The quest for efficient use of limited water resource has also led to the assessment of relative merits of different irrigation methods. It has been found that on equal quantity of water application, the yield of sprinkler irrigated wheat was 43.01 q/ha which was 36.5 and 33 per cent higher than the yields under border strip and check basin respectively.

However, any significant effect of flat, furrow and sprinkler irrigation on cotton (var,·216 F, 414 and Andrews) did not find but that soil water tension has a greater relevance.

Table 14

Production for a Given Quantity of Water Used over Different Areas

Crop	Water use (cm)	On area (ha)	Production (q)
Wheat	84	1	55
		3	91
Bajra (HB 3)	25	1	42
		4	103
Jowar (CSH 1)	28	1	44
		2.5	64
Sunflower (EC 68414)	50	1	16
		2	26
Mustard (Haryana 1)	25	1	11
		1.5	20

In this regard trickle system of irrigation, particularly in case of fruit and vegetable crops, presents considerable promise for optimising the production with limited water resources. Evidences from contemporary literature indicate a number of advantages of trickle irrigation such as (a) 30-50 per cent water economy may be achieved as compared to furrow irrigation permitting a diversion of the limited water over a larger area, (b) produce is improved, (c) yield is increased as compared to both, furrow and sprinkler irrigation, (d) water with excess of salts, as often encountered in arid areas, may be used for irrigation, and (e) nutrients may be directly applied through drip system.

Investigations undertaken here indicated (Table 15) that the yield and efficiency of water use of a number of crops like bottlegourd, ridgegourd, roundgourd and watermelon may be increased to different degrees by the adoption of trickle irrigation method as compared to those obtained under furrow and sprinkler irrigation.

While it is generally acknowledged the trickle system is well suited for high value vegetable and plantation crops, there is a dearth of data on the economic of returns from specific crops in the context of regional soil environment conditions. In the current operation research programme of this institute in Dahijar village near Jodhpur, about 0.6 ha of area has been brought under potato

(1 metre spacing where each lateral feeding double rows of plants spaced 15 to 20 cm apart over a length of 30 to 50 metres. Individual drippers on the laterals were 50 cm apart), where the initial investment has been estimated by Panjab Singh of this institute to be roughly Rs. 18,000 per hectare (excluding pump installation). The cost has been kept low by the fabrication of the drippers (ca. Rs. 0.45 per dripper) in this institute from high density polyethylene. The initial cost, however, has been relatively less (about Rs. 6,000 per ha) in case of orchards of pomegranate and citrus in ORP area where plants were located at distances of six m or more. The whole economics, however, have to be viewed from the profits gained over time. Efforts are also being made here to reduce the overall cost by reducing the number of laterals through adjustment of planting pattern and system design.

Table 15

Yield and Water Use Efficiency of Vegetable Crops Under Various Methods of Irrigation

Crop	Irrigation Method	Water use (cm)	Yield (tonnes/ha)	Yield of water use (q/cm)
	Furrow	84	38.01	4.5
Bottlegound	Sprinkler	84	38.60	4.6
	Trickle	74	55.79	7.5
	Furrow	84	10.74	1.3
Ridgegourd	Sprinkler	84	9.96	1.2
	Trickle	74	12.03	1.6
	Furrow	80	29.47	3.7
Roundgourd	Sprinkler daily	80	30.84	3.8
	Sprinkler per five-day cycle	80	40.62	4.2
	Trickle	80	40.62	5.1
	Furrow	80	67.24	8.4
	Sprinkler daily	80	68.76	8.6
Watermelon	Sprinkler per five-day cycle	80	74.64	9.3
	Trickle	80	82.33	10.3

The possibilities of the use of saline water through trickle irrigation make this method more suited to arid areas where ground water is often saline. In certain experiments undertaken here in the agronomy section, the yield of potato was about 26 tonnes/ha with water having 3 millimhos conductivity and about 14 tonnes/ha with water having 10 millimhos conductivity. Under furrow irrigation with sweet water the yield was 19 tonnes/ha and that under the trickle system was 31 tonnes/ha. It, therefore, seems that the trickle system avoids detrimental accumulation of salts in the vicnity of the roots and thus water having a high salt content can also be used under this system.

Concluding Remarks

The foregoing account of diverse technologies provide ample confidence for optimisation of production from dry regions. However, the crux of the problem is the effective diffusion of these technologies in rural areas. Rational land use in reckoning with the land capability is another dimension of the problem which warrants special reckoning in these areas for conservation of the fragile eco-system.

20

Forestry in the Western Rajasthan

C.M. MATHUR

Western Rajasthan is the synonym of Rajasthan desert or Thar desert. It occupies an area of 1.96 lakh sq. km. and encompasses 11 districts of Rajasthan namely Ganganagar, Bikaner, Barmer, Jaisalmer, Jodhpur, Churu, Nagaur, Jalore, Jhunjhunu, Sikar and Pali. The total coverage of the area is approximately three-fifths of the state.

The natural vegetation of this tract is quite sparse with limited number of species. The forest forms a negligible proportion of total geographical area and is even less than two per cent. The climate is of extreme nature and is characterised by high solar radiation, extreme variations of diurnal and seasonal temperature and erratic rainfall with marked variation from year to year. Usually drought visits this region once in a period of three years. The atmospheric humidity remains very low for the greater part of the year, except the rainy season. The wind velocity remains very strong, particularly in summer months, with frequent occurrence of dust storms. As a characteristic of desertic soil, the soil of this region is immature and impoverished. It is characterised by poor water holding capacity, very low in organic matter, weaker in exchangeable bases, and low in nitrogen and phosphate content. The clay content of soil is very less. Soil is usually sandy to sandy loam type except in some areas of the eastern part of the tract. Thus inhospitable climatic condition coupled with adverse edaphic state plays a dominant role in shaping the vegetation status of the region.

It will not suffice, if, we ignore the dominant behaviour of biotic factors operating in this region. Unlike other deserts of the world, the desert of Rajasthan supports a human population of 8.84 million with almost its double bovine population to the tune of 16.28 million. The sparse and limited vegetation is variously exploited by man through cutting, lopping, felling, stripping and even scrapping the ground vegetation for his immediate need. The constant mishandling and destructive activities on vegetation result in modification of the plant shape into sturted, dwarf, badly mutilated, crooked or prostrate forms. The natural grasslands are mostly overgrased, as a result, make a way for growth of unpalatable and less nutritive herbaceous vegetation. Besides grazing, browsing, trampling, treading and trudging by the domesticated animals, the natural vegetations are subjected to other biotic stresses. The birds, rodents and other herbivores do a considerable damage to the vegetation at various stages of its development. The damage done by termite and locust pests to the flora of this region must not be under-estimated.

The vegetation of the region is xerophytic scrub type falling mainly under tropical thorn forests. The main tree species found in this zone are: *Prosopis cineraria, Acacia Senegal, Tecomella undulate, Salvadora porsica, Savadora oleoides, Zizyphus mauritiana, Maytenus emarginatus, Balanties aegyptica* etc. The shrub community generally consists of *Calligonum polygenoides, Capparis deciduas, Acacia jacquemontii, Zizyphus mummularia, Crotalaria burhia, Aerva pseudotomentosa, Leptadtnia pyrotechnica* and *Calotropis procera.* The comman grasses met with are *Lasiurus sindicus, cenchrus ciliaris, Cenchrus setigerus, Cenchrus biflorus, Dichanthiun annulatum Eleusine compressa, Dactyloctenium sindicum, Panicum spp, Aristida sp,* and *Sporebolus diander* etc.

The principal tree species of Rajasthan–*Anogeissus pendula-occurs* in this region to a limited extent in the natural forests mostly on rocky areas in Pali, Jalore, Sikar and Jhunjhunu districts. *Boswellia serrata* is usually met with in the higher ridges where *Anogeissus pendula* is found in lower in Pali district. According to the recent administrative report of forest department, forest areas in western Rajasthan are as follows (in sq.kms.):

Sikar	662
Jhunjhunu	435
Nagaur	164
Pali	1030
Jalore	326
Barmer	282
Jaisalmer	77
Churu	73
Bikaner	155
Jodhpur	92
Hanumangarh (Ganganagar District)*	102
Suratgarh (Ganganagar District)*	134
Chhatargarh (Bikaner District)*	92

* These areas are mostly new plantation patches, not existing forest.

The above figures show the areas under forest department are very meagre and productivity is also very low. It hardly delivers any goods and services to the community. Realising the great famine of fuel and fodder, the National Commission on Agriculture (NCA) suggested, in its report of 1976, for massive afforestation and pasture development works in this region to augment the productivity. Based on these recommendations of NCA and expertise gained through the scientific technologies evolved by Central Arid Zone Research Institute, Jodhpur, a massive afforestation and pasture development work has been initiated in this region by the state forest department under desert development programme from 1978. Prior to this, in the irrigated areas of Rajasthan Canal in Bikaner and Ganganagar districts, various afforestation and pasture development works were taken up under Rajasthan Canal Project (RCP) Schemes. Thus, to organise and monitor the huge plantation schemes, both under DDP and RCP, a directorate of desert afforestation and pasture development has been set up since 1978. In order to concentrate the activities, three circles have been created at Jodhpur, Sikar and Bikaner respectively and fifteen divisions are functioning in these three circles to give an impetus to the gigantic task undertaken ever since in this region. While planning and monitoring the job, various operational and financial constraints are clearly envisaged alongwith natural constraints and remedial measures are also taken to overcome these constraints as far as possible. However, its greater success depends upon the public participation and their dedicated involvement in the work. As we have seen the forest areas are very much limited and these works are taken up in mostly waste lands under revenue department or on community land under village panchayat and even to some extent in the private holdings of farmers.

The various schemes under DDP and RCP are described as follow.

Desert Development Programme

Under the programme, following forestry schemes have been taken up:

1. Farm forestry
2. Silvi-pastoral plantation
3. Sand-dune stabilisation
4. Shelterbelt-cum-roadside plantation
5. Fodder bank

Working Details of Schemes

1. Farm Forestry

Shortage of fuel wood and its high selling prices in cities have resulted in systematic clearing of the tree growth from the agricultural fields aggravating the process of desertification in this region. Merits of farm forestry consisting of planting trees on the farm boundaries and in the farm itself are quite obvious in as much as the plants would provide wind breaks and meet nearly the farmer's entire fuel requirement besides providing top feed for his camels, sheeps, goats and cattle. In spite of the obvious benefits, farm forestry, as a programme, has not succeded since the forest department limited itself to distribution of sapling to the farmers and leaving the rest to them. In the hostile climate of this region, survival has been insignificant and this has naturally adversely affected the programme. The present scheme is, therefore, being implemented in a few villages in each district where individual willing farmers are being identified by the staff of the forest department in collaboration with the officials of the block and local elected sarpanchas of the villages.

The selection of species for individual farmers is being made generally from the *Prosopis cineraria, Acacianilotica var indica, Zizyphus maruritiana, Acaeia tortillis Dalbergia sissoo, Laucaena leucocephala, Cordia spp.* and *Morus alba,* species depending on the soil conditions, rainfall pattern and availability of water for irrigating the plants. These species would give much needed fuel wood, small timber and fodder to the farmers besides providing organic contents.

The planning is being got done by the extension agency of the forest departmental through the farmer himself who also does protection, watering, fencing, maintenance etc. for which labour charges are paid by the department in the first year in two instalments, first at the time of planting in July-August at the rate of Re.1 per plant and then again in February- March at the same rate for the surviving plants. The farmers will be paid maintenance cost for a further period of two years at the rate of Re.1 per plant per year on the basis of surviving plants in the month of February-March of the year. It is envisaged that after seeing the results of planning in the selected villages and benefits accrued from them, the farmers will take up such plantings on their own in future after obtaining their requirement of saplings from nurseries. During the period of Sixth Plan a target of 20 lakh plants has been kept in this scheme for distribution to the farmers for planning.

2. Silvi-pastoral Plantation

Under this programme, during the plan period, about 11,600 hectares are to be developed. The technique consists of developing one-third of the land which may be either forest land or government wasteland for planting of tree spp. at interval of 5m × 5m and developing the remaining two-third area as grassland where ridges and furrows are formed by ploughing and perennial grass spp. like *Cenchrus ciliaris, Cenchrus setigerus* or *Lasiurus sindicus* are sown on ridges at 30 cm apart. Ridge to ridge distance is anything between 90 and 120 cm. The grass patch is opened for grazing as per the carrying capacity of the site after second seedfall while tree plot is protected for five years. In between two plots of grass and tree land, a ditch fence is provided to protect plants. The cultural operations are to be continued for four years in the tree plot. Normally species recommended for planting are *Acacia tortilis, Zizvphus mauritiana, Prosopis chilensis* and fodder species chosen are *Leucaena leucocephala, Prosopis chineraria, Dichrostachys nutnas, Ailanthus excelsa* etc. During the plant period, about 5,000 hectares area is to be developed as village fuel wood and fodder plantation.

3. Sand-dune Stabilisation

Under this programme, during the period of Sixth Plan, about 7,000 hectares of active and semi-stabilised sand dunes are to be afforested. The technique consists of (i) protection against biotic interference through fencing with angle iron posts and barbed wire, (ii) the treatment of shifting sand dunes by fixing barriers in parallel strips or in a chess board design using local shrub material starting from the crest down to the heel of the dune to protect the seedlings from burial or exposure by blowing of sand, (iii) revegetating by transplanting seedlings of suitable tree species at 5m × 5m spacing, (iv) planting grass seedlings and sowing of grass seeds on the leeward side of brushwood barriers at 30 cm intervals, and (v) continuous after care for a period of four years thereafter. For sand dune stabilisation, major tree species chosen is *Acacia tortilis* with other spp. comprising *Prosopis cineraria, Acacia nubica, Zizyphus mauritiana, Calligonum polygonoides, Acacia senegal* and *Dichrostachys nutans*. Among the grasses, *Lashirus sindicus* and *cenchrus* species are used.

4. Shelterbelt-cum-Roadside Plantation

Under this programme, during the plan period about 10,200 row kilometers are to be planted along either side of the road. The technique consists of (i) protection against biotic interference through fencing the strip with angle iron posts or wooden posts subject to availability and barbed wire followed by, (ii) planting of three staggered rows of trees on each side of the road where plant to plant distance within a row is kept 5 m and row to row distance is kept 3 m, thus 200 plants in a row kilometer, and (iii) continuous after care for a period of four years which includes watering, weeding, hoeing and other cultural operations. Saplings of trees like *Azadiraceta indica, Albizzia leboek, Acacia tortills, Cassia siamea,*

Eucalyptus spp. are chosen for the shade row, i.e., innermost row, *Pongamia pinnata, Acacia tortilis, Tamarix* articulate etc. for the second row and *Acacia tortilis, Acacia nilotina,* Zizvphus *mauritiana, Prosopis chilensis Parkinsconia aculeais* for the last row.

5. Fodder Bank

As an insurance against recurrent fodder famines in the desert region, fodder banks are being created in Jaisalmer, Pali and Barmer districts, where grass is collected from the grass reserves and arrangements made for its storage and preservation. About 2.5 lakh quintals of grass will be collected under this scheme during the plan period. Since the rainfall in this region is low, storage is done in the open in hay stacks which are thatched properly before the onset of monsoon. It is proposed that hay should normally be kept stored for two years but may be disposed off in the third year if not used up in the meantime to meet the requirement during scarcity. Thus, there will be a stock of two years production of hay in the fodder bank.

Physical Achievements

The physical targets achieved during 1979-80 and 1980-81 under different schemes of DDP are given below:

Name of scheme	Unit	Physical targets achieved
Farm forestry	No. of plants in lakhs	12.08
Silvi pastoral plantation	Hectares	8,544
Village fuelwood and fodder plantation	Hectares	2,842
Sand dune stabilisation	Hectares	8,066
Shelterbelt plantation	Row kilometer	4,304
Fodder banking (grass collection)	Quintals	71,000

In spite of severe drought conditions prevailing both in 1979 and 1980, results of plantations are generally very good at most of the sites. Overall survival of plants in plantations raised in the most arid district of Jaisalmer has been over 75 per cent so far. One and a half year old *Acacia tortilis* plants on sand dunes have attained maximum height of 2.5 m. while the average height is 1.5 m.

Financial Outlay

The proposed financial outlay under the Desert Development Programme for the various forestry schemes mentioned above for the period of Sixth Five-Year Plan has been kept as Rs. 18.50 crores only.

Rajasthan Canal Project

The afforestation programme in Rajasthan Canal Project area which envisages protection of canals, roads and farmlands from shifting sand dunes and supply of fuel wood, timber and fodder, is being implemented since 1974-75 with the assistance of the World Bank within the project area of 2.00 lakh hectares (Phase I of Stage I). For the remaining 2.46 lakh hectares (Phase II Stage I) advance action has been initiated last year (1979-80) for raising plantations in this year (1980-81). The various schemes being implemented are:

1. Shelterbelt plantations along canals
2. Shelterbelt plantation along roads
3. Fuelwood plantations near villages
4. Sand dune stabilisation cum pasture development in un-commanded areas of project area
5. Farm forestry programme through farmers (department of supply plants and know-how)

Shelterbelt Plantations along Canals

Tall plants of suitable species such as *Dalbergia aissoo, Acacia nilotica, Bucalyptus spp.* are planted along canals in burrow-pit areas and such other patches where slow irrigation is possible. In un-command areas and on sand dunes, *Cacia tortilis* and *Tamarix articulate* are planted. Rows of the plants are kept 5 meters apart and distance between plants and rows is 3 meters; thus giving 666 plants per hectare. The plantations are done in the quniquinnial pattern.

Dalbergia sissoo (siham) and *Acacia aittica* (babool) are expected to attain a height of about 20 meters and a diameter of 50 centimeters at the end of its rotation period of thirty years. *Acacia tortilis, Eucalyptu and Tamarix articulate* will have a rotation period of ten years. The internal rate of return of this scheme works out to be 12 per cent. This is apart from protective aspects to shelterbelts, which will prevent direct damage to the canals and roads by wind blown sands.

The physical targets of 5,500 hectares under this scheme in phase I during the project period had been achieved. In phase II, 3,000 hectares have to be planted in the project period of five years out of which 1,000 hectares have been planted in 1980-81. About 1,200 hectares of area has also been planted along canal in stage II during 1980-81.

Shelterbelt Plantation along Roads

Tall plants of suitable species such as *Dalbergia sissoo, Acacia nilotica, Eucalyptus, Acacia tortilis* and *Tamarix articulate* are planted on the waste lands along the roads. Three rows of such plants are raised on either side of the road. The plants are planted at a spacing of 5 × 3 meters. Wherever the roads are passing through unstabilised sand dunes, planting is done only on the windward side of the roads and six rows of plants are raised. These are planted with *Tamarix articulate* and *acacia tortilis.* In such stretches leeward side remains un-planted to dump blown sand for cleaning subsequently. These tree belts are irrigated either by flow directly from canal or by tankers.

The internal rate of return of this scheme works out to be 9 per cent. The beneficial effects of raising roadside shelterbelts are obvious wherever plantations have been raised.

The target of raising plantation on 700 hectares of area under this scheme during the project period has been achieved. In phase II, advance action (fencing, soil work) on a length of 900 row km. of road has been completed where planting work was done this year (1980-81). Total length of 3,000 row kilometers has been proposed to be planted in phase II area.

Fuelwood Plantations near Villages

In Rajasthan Canai area, the fuel needs of the people are met from naturally growing khejra (*Prosopis spicigera*) trees and phog (*Colligonum polygonoides*) bushes. With the advance of irrigation and settlement of the area, there is spurt in population growth and as such this natural source of fuelwood is depleting fast. The rural population will therefore in course of time meet its requirement of fuel by burning cow-dung and agriculture wastes. To divert the cow-dung to be used for manure and to conserve the natural vegetation cover, adequate supplies of the fuelwood at reasonable prices have to be ensured. As such a block of 12 hectares of irrigated land is earmarked and reserved in each village, which is being planted up for raising village fuelwood plantation with irrigation for the benefit of the village concerned.

The species being raised in such plantation are *Dalbergia sisoo, Acacia nilotica* and *Eucalyptus*. The plants are planted at a spacing of 5 × 3 meters, thus giving 666 plants per hectare. The internal rate of return for this scheme works out to 11 per cent. Besides, these plantation will improve the landscape and provide recreational facilities and ameliorate the micro-climate.

The target of 1800 hectares under this scheme during the project period has been achieved. In phase II, 400 hectares have been planted in 1980-81. Total area to be planted in phase II is 1,850 hectares.

Sand-dune Stabilisation cum Pasture Development

Forty per cent of land of the Rajasthan Canal Project will remain un-commanded by irrigation. This is mostly in the form of sand dunes which are a constant menace to the canal system and irrigated farm land. Thus to stabilise these dunes and also to meet the fodder requirements of cattle breeders, a programme of sand dune stabilisation and pasture development is being implemented. Developed areas of over two years are being opened for limited grazing of cattle according to their carrying capacity from November to March each year.

Supply of Plants to the Farmers

It is proposed to raise and supply 70,00,000 seedlings to the farmers for planting along water courses in phase II of stage I of the World Bank Project.

Physical Achievements

Even before the implementation of the above mentioned schemes under World Bank Project since 1974-75, plantation work in the Rajasthan Canal Project area was initiated in 1962-63 under the normal budget provisions of the department. Later some areas were also planted under Drought-Prone Area Programme and World

Food Programme. Total physical targets achieved under different programmes in this area since 1962-63 to 1980-81 are given below:

Name of scheme	Unit	Physical targets achieved
Canal side plantation	Hectares	12,560
Roadside plantation	Hectares	1,670
Village fuelwood plantation	Hectares	3,570
Sand dune stabilisation cum pasture development	Hectares	62,000
Supply of plants to farmers (schemes initiated in 1980-81)	No. of plants in lakhs	6.5

Financial Outlay

The proposed financial outlay for all the above mentioned schemes under phase II of stage I in the Rajasthan Canal Project area for the project period of five years has been kept as Rs. 7.7 crores.

Conclusion

Forestry operation in the desert region outside Rajasthan Canal Project area is quite a challenging task because there are many odds in the way to it. The magnitude of work taken up is very meagre compared to the vast fields lying barren in this area. But certainly with these works, an infrastructural base has been built up and further development can be done in the area. People's participation is very vital for all these works, otherwise no programme will be successful at all. The present work, though in a small scale has created a suitable atmosphere of confidence both for the working staff as well as the public at large.

IV

Famines and Droughts

21

Prediction of Drought in Arid Rajasthan Folklore

SANJAY SHARMA

Rajasthan is the second largest state of the Indian Union. Situated in the west of the country, it occupies 10.43 per cent and 5.13 per cent of the total land area and population of the country respectively. Geographically, the state is divided into two distinct parts by the Aravalli range of hills which is one of the oldest mountain systems in the world. The range stretches diagonally across the state in the southwest to northeast direction. On the west and north lie water-deficient, arid plains and the ever-shifting sand dunes of Jodhpur, Jaisalmer and Bikaner. This sandy region in the west of Aravalli range is popularly known as the Great Indian Desert, which is a part of the world's largest Thar Desert. On the east and southeast of Aravalli are the forests and greener plains of black loam trenched by the running streams and rivers.

Agriculture and animal husbandry are the main sources of livelihood of the people. But, being largely water-deficient and lacking in vegetative coverage, people have been facing grave hardships, which is further accentuated by vagaries of monsoon. Failure of monsoon has often resulted in severe droughts and subsequent famines.

Drought and Famine

Drought is a natural phenomenon that occurs fairly periodically all over the world. Therefore, it has become a central problem of concern. In the last decade Ethopia, Somalia, Australia, India and many other countries were severely affected by drought.

Literally speaking, drought means a considerably lower rainfall from normal. Indian Meteorological Department has defined it as "a situation occurring in any one area when the rainfall is less than 75 per cent of the normal; and when deficiency of rainfall is about 50 per cent of the normal, it is termed a severe drought".[1]

Thus, failure of monsoon in an area, as in case of Rajasthan, where more than two-thirds of the area under cultivation is rainfed and where irrigation system is largely poor and not sufficient, often creates four types of hardships viz., critical fall in employment, food, fodder and drinking water for both human and cattle population.

Famine, on the other hand, may be defined as a state of affairs in which people of a region suffer from acute shortage of either food, fodder and water or any one of them. Famines are thus classified as *Annakal* (grain famine), *Jalakal* (scarcity of water), *Trinakal* (scarcity of fodder) and *Trikal* (scarcity of grain, water and fodder). Such conditions may naturally occur due to drought, floods, excessive rainfall, insect-pests, cold waves etc., which by producing adverse climatic conditions or in other ways cause crop failures and give rise to famine conditions.

Rajasthan, owing to its geographical situation and climatic conditions, has always been subject to drought and subsequent famines. Due to unpredictability of monsoon, the people of the state, on an average, have to face one drought year after every three years. In Marwar, people generally expect one lean year in every three years and one famine year in every eight years.[2] There is a couplet which tells us about severity and chief visiting areas of famines in western Rajasthan:

पग पूंगल धड़ मेड़ता, उदरज बीकानेर।
भूलो चूको जोधपुर, ठाँवो जैसलमेर।।[3]

[My (famine's) feet remain in Pungal (a district of Bikaner), head in Merta (*a paragana* of Jodhpur) and belly in Bikaner. At times I do visit Jodhpur, but my permanent abode is in Jaisalmer.]

Weather Prediction Indicators and Folklore

Culture, in its broadest sense, refers to that part of total repertoire of human action (and its products), which is socially opposed to genetically transmitted. Taylor,[4] while extensively defining the term, said: "Culture of civilization is that complex whole which includes knowledge, belief, art, moral, law, custom, and other capabilities and habits acquired by man as a member of society." Thus, culture is a social heritage; it is the gift of society to an individual. Radcliffe Brown[5] regards culture as cultivation, the process of handling down and acquiring traditions, as a result of which a society is perpetuated. While, for Malinowski[6], culture is essentially a response to human needs.

Keeping in view these definitions, an attempt has been made in this paper to suggest how the people of this arid region have, over the years, cultivated and passed over the indigenous knowledge in predicting the onset of monsoon, drought and famine, obviously a response to meet human needs.

Meeting the hardships of adverse climate, people of this land have been watching the natural phenomena occurring just before and during drought and famines. Through their observations they have arrived at a relationship between such phenomena and the likelihood of an impeding drought and famine. Thus, studying the changes of weather with apparent indicators like qualitative changes in climate, vegetation, animal and human behaviour, these people have developed a meteorological science of their own. This science, based on empirical generalisation through observations, is not entirely unreliable. This indigenous knowledge, developed and verified through ages, is verbally passed on from generation to generation in the form of verses, sayings, couplets etc. Thus, villagers' folk culture depicts the indigenous knowledge and beliefs about the nature and its peculiarities.

Naturally occurring indicators like wind, direction, temperature, cloud, month, date and day, zodiac periods, moon and stars and their movements, were and still are taken to be the most reliable (though often untrue) indicators for predicting the onset or failure of monsoon, i.e., a probability of impeding drought and famine. There are numerous verses and sayings common among people that narrate said relationship between climatic elements and drought and famine or the prospects of good or bad yield of crops.

Following are a few of such sayings and verses that narrate the predictability of monsoon or famine:

आसाड़े धुर अष्ठमी, चन्द उगन्तो जोय।
कालो वे तो करवरो, धोलो वे तो सुगाल।।
जे चन्दो निरमल होवे, तो पड़े अचित्यो काल।।[7]

[One the eighth day of dark half of *Asadh*, watch the rising of the moon. If it rises in black clouds the year will be an ordinary one. If in white clouds the year will be a good one. If the sky is cloudless there will be an unimaginable famine.]

सावण वद एकादसो, तीन नरवत्तर जोय।
(किरत्का) होवे तो करवरी, रोहन होय सुणाल।।
टुक यक आवे मिररालो, तो पड़े अचित्यो काल।[8]

[If on eleventh day of the dark half of *Sawan*, the moon be present in the following three *Nakshatras*, the effect will be as follows: If in *Kirtika* (Pleides), the year will be an ordinary one, if in *Rohini* (11 Debran), the year will be a prosperous one, and if in *Miragsaro* (Orion), it will be a famine year.]

परभाते गेह ड़ंबरा, साजे सोला बाब।
डंक कहे हे बादली, काला तण सुभाव।।[9]

[Clouds flying in the morning and cold winds blowing in the evening, says Dank, are the sure signs of famine.]

परभाते गेह ड़ंबरा, दोफारा तंपत।
रात तारा निरमला, चेला करो गछत।।[10]

[Flying clouds seen in the morning, very hot in the mid-day, night cloudless with shining stars, it is time to leave the native place, i.e., these are the sure signs of famine.]

चैत मास उजाले पंख, नव दिन बीज लुकाई रख।
आठम नम नीरज कर जोय, जा बरसे जा दुरभख होय।।[11]

[During the bright half of *Chait*, let there be no flash of lightening for the first nine days; especially on the eighth and ninth days, wherever it rains in those days, it will be a sure famine.]

सावन में सुरयो चहलै भद्रावे परवाई।
आसौज में पिछवा चलै भर भर गड़ा लायै।।[12]

[If the north-western wind blows in *Sawan*, the eastern wind blows in *Bhadaun*, or the western wind in *Asoj*, they bring carts full of grain (i.e., predicting good monsoon.]

जैठ बीती पहले परवा, खतक अंबर हरी।
असाढ़ सावन खेत सूखो, भादर हुए बिरखा करै।।[13]

[Thunder on the first day after the end of *Jeth* means two dry months and no rain before *Bhadaun*.]

तीतर पंखी बदली बिधवा काजल रेख।
ओ बरसे ओ धर करै, यामें न मेख।।[14]

[Clouds with wings like a partridge and a widow with *kajal* in her eyes, without doubt it is a sure indication that the former will bring rains and the latter will remarry.]

उगन्तरो मछालो, आठनवतरो मोख।
डंक कहे हे भडली नदियां अहदसी गोख।।[15]

[Rainbow at sunrise, red rays of the setting sun. O Bhaddli, says Dank, augurs over flooding of rivers.]

चैत मास ने पख अंधियारा।
अहै चौदस दो दिन सारा।।
जिन दिस बादल, जिन दिस मेह।
जिन दिस निरमल जिन दिस खै।।[16]

[During the dark-half of *Chait*, especially on the eighth and fourteenth day, towards the part of the country where clouds appear, there will be good rain and in the cloudless direction, only sand will blow, i.e., there will be no rain.]

Similarly, a study of folklores reveals that people also used behavioural changes of animals (jackal, she camel, dogs, rats, sheep and goats etc.) and plants and trees (like *khejri, neem, kair* and sewan-grass etc.) in predicting the nature of forthcoming weather. Since animals and birds are more sensitive to physical changes than humans, people watch their activities for sign of strange behaviour for an early precognition of ensuing weather. For example:

सावन सुरंगी खेजड़ी, कातो बिरंगा खेत।
'सावन बिरंगी खेजड़ी, काती सुरंगा खेत।।[17]

[If the *khejri* is colourful in *Savan*, the fields are colourless in *Kati*. If the *khejri* is colourless in *Savan*, the fields will be colourful in *Kati*.]

दिन में सयाल सबद जो करे।
निश्चय ही काल हलसाल पड़े।।[18]

[When the jackal howls during the day, a great famine is certain.]

Apart from predicting the onset of famine, many of the verses and sayings reveal the extent of severity of famines that has often resulted in migration, deaths due to starvation and thirst, and generation of many social problems like selling of one's own child, separation of married couples and even state of anarchy etc. The following verses/couplets indicate the dire socio-economic consequences of drought and famines:

सावन पहले पख में, जो तिथि उणी जाय।
कइ एक कइ एक देस में, टाबर बेचे माय।।[19]

[If there be a substractive *tithi* during the dark half of *Sawan*, somewhere in the country mothers will have to sell their children (to get food), so severe will be the famine.]

सावण पहली पंचमी, जो बाजे बहु बाय।
काल पड़े सहु देस में, मिनख मिनख न खाए ।।[20]

[If on the fifth day of *Sawan*, the winds blow strong, a famine will prevail throughout the land and man will eat man.]

बरसे भरणी, छोड़े परणी।[21]

[If it rains in *Bharani* (from 25th April to 8th May), a husband has to abandon his wife (i.e., the famine expected is of such a severity that the husband will not be able to support even his partner-in-life).]

मिरगा बाव न बाजियो, रोहण तपो न जेठ।
के नै बांधों झूंपड़ों, बैठो बड़ले हेठ।।[22]

[If no winds blow in *Mirga* (6th to 19th June), if it be not very hot in *Rohini* (from 2nd May to 5th June) and *Jeth* (from 2st June to 3rd July), why should you build a hut, we shall have to live under a Banyan tree or wander (will be homeless on account of famine and consequent emigration).]

नाड़ा ट़ाकंण बलद बिकावण, मत बाजे तु आधे सावण।[23]

[O! You, *nara tankan* (south-easterly wind), don't blow in the middle of *Savan* (August), it will make me sell my bullocks (i.e., famine will result in selling of one's (farmer's) bullocks).]

माह मंगल जेठ रवी, भा–दखे सन होय।
डंक कहे हे भडली, बिरला जीवो कोय।।[24]

[If there are five Tuesdays in the month of *Magh* (January-February), five Sundays in *Jeth* (May-June) and five Saturday in

Bhadaun (August-September), Dank says to Bhadli, so severe will be the famine that few will survive.]

आदो बलो उलांगियों, छोड़ी घर री आस।[25]

[When the Aravalli hills are once crossed, all hopes of returning home are lost (i.e., formerly whenever the famine stricken emigrated to other provinces, they left their motherland with little hope of returning home).]

सावन पहली पंचमी, मेह न मांड़े आल।
पीव पड़सो मालवै, मैं जांसा मोसाल।।[26]

[On the fifth day of dark half of *Savan*, if rain does not make a beginning, dear husband, you go to Malwa and I will go to my father's house. Thus, emigration had been the main source of relief in times of scarcity and famine.]

In the present century, science has taken many strides in meteorology. Highly sophisticated satellites and computers can accurately predict the weather of any remote area. Meteorologists have found global interconnection of weather and climatic phenomena. The severe drought of 1987 in India was, thus, attributed to the teleconnections of warm currents of Peru and low atmospheric pressure at Port Darwin. This phenomenon was called 'El-Nino Effect.'

Since independence, the construction of Indira Gandhi Canal and tubewells have, to an extent, satisfied the need of water in different parts of the arid Rajasthan. With greater emphasis on dryland farming and with the increased use of fertilisers and high yielding varieties of short duration, drought resistant crops; the agricultural productivity of the region has increased. As a result the dire severity of drought has greatly been reduced but not completely eroded, as monsoon rains still subserve the subsistence agricultural needs of this region.

Ignorant of advanced meteorological science and newer high yielding agricultural practices, these largely poor and illiterate people of arid Rajasthan still rely heavily on their own indigenous knowledge of drought and famine predictions. Even now, this generation-old indigenous knowledge, culturally passed on to the next generation in the form of folklores, couplets and verses, forewarns these people about the likelihood of impeding drought or monsoon failure.

NOTES

1. *Farmers' Journal*, A.J. Publication, September, 1987, p.4.
2. Gahlot, J.S., *The History of Rajputana*, Vol. II, Jodhpur, 1937, p. 122 (RUCL).
3. Gahlot, J.S., *Rajasthan Before Second World War*, Rajasthani Granthagar, Jodhpur, 1986, p. 124.
4. Taylor, E.B., *Primitive Culture*, Vol. I, London, 1871.
5. Radcliffe-Brown, *A Natural Science of Society*, 1957.
6. Malinowski, B., *A Scientific Theory of Culture*, University of North Carolina Press, 1947.
7. Gahlot, J.S., *Rajasthani Krishi Kahavaten*, Jodhpur, 1941, p.45.
8. Ibid., p. 24.
9. Ibid., p. 9.
10. Ibid., p. 10.
11. Ibid., p. 20.
12. Bharara, L.P., "Indigenous Knowledge: A Part of Cultural Heritage in Mitigating Drought Effects in Arid Zone of Rajasthan." Paper presented at the National Seminar on "Integrated Socio-ecological Development of Desert", held at CAZRI, Jodhpur, 19-21 March, 1988, p.6.
13. Ibid., p. 7.
14. Ibid.
15. Gahlot, J.S., *Rajasthani Before Second World War*, op. cit., p.124.
16. Ibid., p. 124.
17. Bharara, L.P., op. cit., p.7.
18. Ibid.
19. Gahlot, J.S., *Rajasthan Krishi Kahavaten*, op. cit., p. 25.
20. Ibid., p. 27.
21. Ibid., p. 34.
22. Ibid., p. 36.
23. Ibid., p. 45.
24. Ibid.
25. File No. C/2/16, Vol. I, "Famine Programme Report, Jodhpur Government", 1939-40, p. 4 (BA).
26. Gahlot, J.S., op. cit., p. 24.

22

Famine Code and Its Administration in the Nineteenth Century Rajasthan

KAMALA JAIN

Famine is a natural phenomenon in Rajasthan. According to a Marwari proverb, there is one lean year in three and one famine year in eight. The proverb was also true of the erstwhile state of Marwar which was visited by at least sixteen famines from 1792 to 1900.[1]

The frequency of famines in India increased in the later half of the 19th century due to the East India Company's policy of exploitation. Competition with British machine-made goods,[2] decay of the cottage industries, destruction of Indian shipping, exemption to British merchants from payment of usual customs and transit duties, the Company's monopoly of Indian trade, heavy duties imposed on Indian goods exported to England, payment of fat salaries to British officials, military expenditure on imperial expansion and annual drain of £1,50,00,000 for home charges and various other methods adopted by the British government also crippled India's economy.[3] Lord Lawrence, Viceroy of India, writing in 1864, also admitted the above fact. He observed: "India is on the whole a very poor country, the mass of the population enjoys only a scanty subsistence. No wonder then that the Indian people suffered grievously in the second half of the nineteenth century with the slightest failure of rain and died in thousands."[4]

In such circumstances it was natural for the famines to be more frequent and severe. The same was true of Rajputana also. During the period under study (later half of the 19th century) Rajputana suffered from three major and two severe famines. Famines of 1868-70, 1891-92 and the *Chappanya Kal* of 1899-1900 all these accrued in the later half of the 19th century.[5]

It is worthwhile to note that the famines in the later half of the 19th century were more disastrous in effect in Rajputana than in many other parts of the country which were directly ruled by the British. On account of the absence of perennial rivers, insufficient and irregular monsoon, desert and hilly tracts and the backwardness of the agriculturists, the affliction in Rajputana was much more widespread and severe than in other provinces. Moreover, the incursions of the Marathas had not only sapped the vitality of the people but also emptied the coffers of most princely states of Rajputana. Agricultural, commerce and cottage industry had declined. Before 1858 the British also paid hardly any attention to the economic development of the princely states. They were busy in extending their empire and consolidating their gains. Besides, the British authorities had yet to formulate any clear-cut policy to combat famines in their own areas.[6]

Famine Relief Policy

The development of Famine Relief Policy in Rajputana during the later half of the 19th century can be divided into two periods: first, from 1858 to 1885 and second, from 1885 to 1900. During the first period traditional policies were followed by the rulers of Rajputana while the British government, through the Agent of the Governor General of India, issued guidelines to the princely states from time to time. To combat the famine, the rulers adopted liberal measures. Whenever it was considered essential taxes were remitted, cooked food was distributed, duty on grains was abolished, grains were imported and relief works were opened to provide work to the able bodied. But, no uniform policy was noticeable in the administration of famine relief in the princely states.

In the year 1868-69, there was a 'Trikal' (great scarcity of food, fodder and water) in Rajputana. At that time, the British government in India, in its famous resolution of September 24,

1868, laid down the policy to be followed by the princely states to meet the crisis.[7] In this resolution, the government made it clear that it had no intention to interfere with the famine administration of the princely states, but they were advised to follow the same policy for famine relief as was being followed in British provinces. They were also asked to give relief to the people by starting useful public works, especially those related to communication and irrigation. Suppression of grain robberies, prevention of sale of children, storage and sale of grains so long as it did not interfere with a free market, providing relief in form of food or cash to indigent women and children, abolition of taxes interfering with the free movement of cereals and preventing people from migrating to other areas were other important measures suggested.[8]

The relief measures were taken as suggested by the British Indian government at the Centre. The princely rulers of Rajputana abolished all duties on grains in transit through their territories. Grain was imported by camels from Sind and Gujarat in Rajputana and by carts along the Agra-Ajmer roads in spite of great difficulties.[10] Many other steps were also taken.

Appointment of a Famine Commission

But, during this period, the policy in British India was also one of experimentation based on trial and error method. It was noticed that the problem was not tackled in a planned manner in the princely states. Similar experience of the British government in its own provinces led to the appointment of a Famine Commission by Lord Starchy. The Commission was asked to examine the famine conditions in princely states also. The efforts of the Commission eventually culminated in the drafting of the provisional Famine Code in 1883 and later a "Draft Famine Code for Princely States" in 1885 as a suggestive help to the administration of the princely states.[11]

However, after the issue of the Famine Code for princely states in 1885, the second period commenced in which the British rulers tried to implement the suggestions of the code by introducing preventive measures during the normal days and worked on the guidelines given in the Famine Code during the continuous period of scarcity from 1889 to 1900.

In a nutshell, the main contents of the Draft Famine Code were:

1. A special officer in each district of the state was to be appointed to watch the condition of crop, cattle and people and submit a report once in a month.
2. The above officer is to submit the report to the central officer. The duty of the central officer was to prepare a statement, when a famine was apprehended, about the area, population, roads, food stocks and about the arrangements to be made and suggestions to help that district in scarcity and to inform the ruler of the state about the same.
3. A copy of this statement should be forwarded to the Central government through its Political Agent for information.
4. When a famine was apprehended, an officer known as the Famine Secretary was to be appointed to exercise control in the state. The above mentioned central officer or anybody else could be appointed to this post.
5. Every Nazim or Tehsildar of the district was to be held responsible for the control of arrangements in the times of famine in his district.
6. The central officer was to prepare the statement of relief works required in that part of the state.
7. Public works officer was to prepare and maintain annually a programme of relief works of permanent utility or importance and these works were to be as big as could be supervised by the state officials. If assistance from the Government of India was desired, the Political Agent was to be informed.
8. In case of famine the central officer was to arrange for employment of those who were unable to leave their homes. The types of works suggested included repair of tanks, embankment of canals, cleaning of drains, fields, stone quarries, etc.

These measures were introduced to prevent emigration of the population for want of works. Likewise, the Famine Code suggested about the medical arrangements, payment to labourers, task works, rules regarding payments of new arrivals, leave tickets, civil agency works (under the control of Nazims and Tehsildars), village inspection, arrangement of poor houses, wages and rations, duties of police department, etc.

Implementation of the Code

After introducing the Draft Famine Code for Rajputana, fortunately, there were a few years (from 1884 to 1889) of respite, providing ample opportunity and time to pay serious attention to preventive measures in the light of the experience gained in previous famines and suggestions made by the Famine Code.[13]

Many important irrigation and communication works were started and completed in Rajputana during these good days as a precautionary measure. Good means of communication play an important part in transporting foodgrains rapidly to the deficient areas and helping the sufferers in migrating to surplus territories. From a careful study of the administration reports of the states it is noticed that the authorities were not negligent in this matter. For example, the road from Udaipur to Chittor was completed in 1883-84. It joined the capital of Mewar to an important junction of Rajputana-Malwa railway line. By the construction of a bridge over the Berach river in 1884-85, the most important obstacle on railway line was removed. The Agra-Ahmedabad road in Jodhpur territory was another important public work in the year 1883-84. In Jaipur, a new road was built in the year 1883-84 from Chandpole Gate of the city to the railway station.[14] In Alwar, the construction of Ramagarh-Firojpur road, Umran-Thana Ghazi road and Kathumer-Kehtri railway feeder road were completed. In Kota, the Borde Ghat carisway (2300 feet long) on the Purwan river was completed in 1884-85 at a cost of Rs. 27,778.

In the same way Rs. 25,70,989 were spent in the year 1882 on construction of railways in Rajputana. These figures show a steady progress in the expenditure of public works department of the principal princely states under the Rajputana Agency.

Likewise, many irrigation works were taken in hand and were completed during this period. In Jaipur, major irrigation works that were completed included Tori Sagar, Madho Sagar Band, Mahashi River Project etc.[15] In Kota, works on Durra Nimboda Saran, Chatarpur tank, Bilasra tank in Nizamat Khanpur, feeder to Eklera Sagar tank from Bilasi river on a new project known as Ramgarh Canal were executed in 1889-90. Many more irrigation and communication works were accomplished in Alwar, Jhalawar, Karauli, Bikaner, Marwar, Tonk, Shahpura, Bundi, Mewar and Dholpur states.[16]

The rulers of the princely states also took measures to implement other recommendations of the Famine Code. Lists of the works which were to be undertaken during the time of a famine were prepared in advance. Both officers and jagirdars were instructed to keep themselves ready for starting operations as soon as they observed scarcity conditions. Sufficient stock of foodgrains was kept at various important places.

It is very creditable on the part of the rulers of the princely states of Rajputana that they laid stress on the construction of irrigational works and development of communication during normal years, although the Famine Code did not guide them properly in these spheres.

As a result of the precautions taken by the different princely states, we find that they were ready to meet an emergency. Such an emergency came soon in 1891-92 in four princely states (viz., Marwar, Mewar, Sirohi and Jaisalmer) of Rajputana where the scarcity of 1889-90 turned into a famine. During the famine of 1891-92 and during the partial famines from 1892 to 1995, the administration of these princely states could cope with the situation in a much more efficient manner.

Nevertheless, the machinery was not geared to such perfection as to combat successfully the dreadful situation created by the continued and widespread famines of 1896-98 and 1899-1900. In 1899, a severe famine scourged many parts of India. According to Lord Curzon, it was the most widely extended and most terrible famine known in India history.[17]

The famine of 1899-1900 known as *Chappanya Kal* as it occurred in V. S. 1956 was a *Trikal* in Rajputana. People who suffered in this famine later told their offspring that such an unusually severe type of famine was not known in Rajputana for over a hundred years.[18]

Relief works were not started in Rajputana until September 1899 except in Marwar. The Maharana of Mewar and officials could not at first grasp the gravity of the situation. In September 1899, Martinadale, the Agent to the Governor General, visited Udaipur and impressed upon the Maharana the necessity of an immediate and comprehensive policy of famine relief. Captain Pinhey, the Boundary Settlement Officer, Mewar, was placed in general control of famine operations. Several relief works were opened in the Khalsa area of the state.[19]

By the end of December 1899 the number of workers on relief works had risen to 23,876. The public works department started a sufficient numbers of works to provide employment to a daily average of 69,000 persons in Marwar. The above figures afford a striking proof of the practical nature of the relief works policy and its efficient administration by the state.[20]

In Mewar, the Maharana opened relief works in the areas of jagirdars also. Foodgrain was imported, revenue collection was suspended, takavi loans were given and medical arrangements were made to combat with cholera and malaria.

In the same manner, in all the princely states of Rajputana, famines relief works started and managed accordingly.

In Bikaner and Alwar, the imperial service corps and in Jaipur, Marwar, Bharatpur and other states the public works departments were given the sole control of the relief works under the direct supervision of the controlling authority of the concerned state. In many instance, the relief programme was supplemented by petty works which were managed entirely by a civil agency. But nowhere was there any approach to a system of dual control.[21] In this famine one leading principle that seemed to have been followed in general was that the people were to be saved from starvation at the least possible expense to the state, by employing them on such works as the rulers deemed most remunerative. This policy was treated like a rule that was justified by its results.

Conclusion

A comparative study of relief operations during the famines of 1868-69 and 1899-1900 in Rajputana clearly shows considerable improvement both in the conception of the ruling chiefs as regards their responsibilities towards the people and in the measures adopted by them.[22] Even then, it was true that due to several factors beyond the control of the rulers there was a wide gap between the administrations of famine relief in the princely states of Rajputana and British India.

During the *Chappanya Kal*, despite all these precautions, the administrations of famine relief in British India and in the states of Rajputana revealed many shortcomings in the working of the

Famine Code. As a result of this experience, the Famine Commission of 1898 recommended greater attention for the relief of tribes. To reduce cattle mortality reserve fodder stock was emphasised. The aims and objectives of charitable relief were also laid down which led to the creation of Indian People's Famine Trust for Charitable Relief. The Famine Commission of 1901 also laid great stress on protective and preventive measures by extending facilities of irrigation and railways.

Thus, it can be seen that the Famine Code was a good suggestive help. But, on the other side, constant rise in prices of grains on account of recurring famines and other factors, for instance, the drain of wealth from India, the high rates of land revenue demands, excessive taxation and over-assessments consequent upon the policy of extravagance followed by the Government of India in both civil and military departments, led to the general impoverishment of the subjects of the British Indian states as well as princely states.

On the whole, the rules of Rajputana were generous enough to open the relief works but the management part of famine administration was poor on the part of their officials.

Famine in our own days no more causes starvation. But it causes semi-starvation on account of inevitable spurt in prices of foodgrains, unaccompanied by a proportional increase in wages, forcing the poor to cut down their consumption of foodgrains. Therefore, a correct and long-term famine policy should aim at banishing poverty and unemployment among the masses so that the purchasing power of the poor masses is increased.

NOTES

1. (a) *Imperial Gazetteer of India*, Provincial Series, Rajputana, 1908, 9.61.

   (b) Handley, T.H., *General Medical History of Rajputana*, p. 175.
2. Dutta, R.C., *Economic History of India: Early British Period*, p. 278.
3. Srivastava, H.S., *The History of Indian Famines*, 1858-1918, p. 10.
4. Naroji, *Poverty and Un-British Rule in India,* pp. 38-51.
5. Maloo, K., *The History of Famines in Rajputana* (1858-1900), 1987, p. 233.
6. Ibid., p. 235.
7. (a) N.A.I. Proceedings, Home (Public) Famine, Nos. 72 to 74, September 26, 1868.

(b) Report of the Indian Famine Commission, 1880, Appendix 11, Resolution of the Government of India, P.W.D , Simla, September 24, 1868 and Annexure to the above resolution.

8. (a) N.A.I., Foreign Department Gen. A., March, 1871, Nos. 34-36 Brooke, J.C. Report on the Famine in Rajputana and Ajmer-Merwara, Para 178.

   (b) Mewar Agency Report, 1868-69, p. 50.

9. (a) Brooke, J.C., Report on the Famine in Rajputana and Ajmer-Merwara, para 171.

   (b) Report of the Indian Famine Commission, 1880, Part III, p. 81.

10. Proceedings, Home Department (Public) Famines, No. 178, 26 December, 1868.

11. N.A.I. Foreign Department, Intl. B., February 1897, No. 97-104.

12. Maloo, K., op.cit., p. 236.

13. Report on the Political Administration of the Rajputana States for 1888-89, p.15.

14. Report on the Political Administration of the Rajputana States for the years 1883-84 and 1884-85.

15. Report on the Political Administration of the Rajputana States for 1888-89, p. 15.

16. Maloo, K., op.cit., pp. 96-97.

17. (a) N.A.I. Government of India, Department of Revenue and Agriculture, No. 11, 2943 (Famine), 20 December, 1900.

    (b) N.A.I. Viceroy's Speech in Legislative Council on October 19, 1990.

    (c) Srivastava, H.S., *The History of Indian Famines*, 1858-1918, p. 240.

18. Final report of the operations of the Indian Famine Charitable Relief Fund in Rajputana and Ajmer-Merwara, upto 31 May, 1901, p. 19.

19. Mewar Agency Report, 1899-1900, p. 5.

20. (a) Report on the Famine in the Native States of Rajputana, 1899-1900. p.44.

    (b) Hakikat Bahi, Jodhpur, No. 38, pp.43, 80, 81, and 82 (Rajasthan State Archives, Bikaner)

21. Report on the Famine in the Native States of Rajputana, 1899-1900, p.73.

22. Ibid., p. 109.

23

Famines and Droughts in Rajasthan

J.P. KANAWA

Famines, droughts and scarcities are words with which people of Rajasthan are quite familiar. These have intimate relation with rural life in the state. Droughts and famines are permanent disasters in the rural Rajasthan, as several districts are frequently in the grip of famine with varying intensity. Very often famine assumes the form of a calamity and the Centre and the state have to struggle hard to face the grim situation lakhs of people are affected with during the periods of scarcity and famines. The government suffers heavy financial burden in two ways: firstly, relief expenditure has to be borne to save the people and livestock from natural catastrophe, and secondly, land revenue collection has to be suspended reducing the revenue receipts of the government to a substantial extent. Inhabitants and animals migrate in search of food and water to the neighbouring states which upsets their economic balancing also. More people go below poverty line and face uncertain situation. Thus, famine and drought indicate fragile nature of rural economic of Rajasthan. Such instability needed to be reduced in the years to come.

In the 11th century Rajasthan faced the most severe drought when there was no rain for 12 years in continuation. The deadly drought of the year 1842-43 was nicknamed as "सहसा मूदसा". The

year 1899 was also a year of severe drought which was named as "छप्पनिया अकाल" as it occurred in the year 1956 of Vikram era. From 1981-82 to 2002-03 only three years remained free from famines and scarcity conditions, viz.,1983-84, 1990-91 and 1994-95. In 1987-88, all the then existing 27 districts were affected by famines and a total of 3.2 crore people of 36,252 villages were under drought condition. Likewise, in 1999-2000, 2.6 crore population of 23,406 villages of 26 districts, in 2000-01, 3.30 crore population of 30,583 villages in 31 districts, and in 2001-02, 4.47 crore population of 40,689 villages of all the 32 districts, was in the grip of famine and there was a great loss of cattle wealth in the state.

Famine Zones/Districts

In different years the famine-affected districts differ widely. Even then, the southern part of Rajasthan is frequently visited by famines. There is a couplet in Rajasthani which means: "*That the legs of famine lie in Bikaner, the main body in Marwar and arms in Barmer. It can be traced in Jodhpur after some search and it has its fixed place in Jaisalmer.*"

The National Agriculture Commission has considered 11 districts as desert districts, which occupy 61 per cent area and 40 per cent population of the state. They are: Jaisalmer, Barmer, Jodhpur, Sriganganagar, Nagaur, Churu, Bikaner, Pali, Jalore, Sikar and Jhunjhunu. The desert areas of these districts always remain in the grip of demon-famine. About two-thirds of the total wastelands of the state is found in these 11 districts. There is lack of underground and surface water, as rains are scanty and irregular in these districts. There are sandy storms and soil erosion due to hot winds.

In the recent years famines have taken the form of famines of water to a great extent. This has affected the life of people and animals adversely. The government can remove the shortage of foodgrains more easily than that of water. In fact, famine in recent years has assumed the form of "triple famine", which means the acute shortage of food, fodder and water simultaneously. That makes the situation quite vulnerable and more challenging.

Causes of Famines and Droughts

The recurrent famines in Rajasthan indicate a state of constant struggle between man and nature. The causes of famines are basically natural, but economic, social and political factors are also responsible for the existence of this problem. These are as follows:

Natural or Geographical Reasons

(a) Physical structure, climate etc.: Desert, which is spread over long distances, and where during summer, land and sky, human being and animals, all are victims of scorching heat, cannot get rid of droughts and famines so easily. There are sand dunes everywhere, and constant wind erosion going on in this area.

(b) Scanty rainfall, its irregularity and uncertainty: Failure of monsoon is the root cause of famines and droughts. In the 11 drought-prone districts of Rajasthan, the average annual rainfall is not more than 50 cm. In Jaisalmer, it is about 16 cm only. During the past 100 years, there was rainfall only in 25 years in this region. Thus, lack of rainfall, total absence of rainfall, late rainfall and irregularity of rainfall—all contribute to the emergence of famines and droughts.

Economic Reasons

The problem of famines gets multiplied due to lack of economic development in desert areas. There is lack of infrastructural development in such regions due to long distances and difficult terrains. Due to decline in traditional cottage and village industries, unemployment has increased in rural areas. Agriculture could not develop due to lack of irrigation facilities. Agricultural productivity is also low in desert areas. Small and marginal farmers and landless labourers face situations of both unemployment and underemployment.

Social Reasons

The people from SCs/STs and other backward communities also face economic difficulties during periods of droughts and famines. For fuel trees from jungles have been cut down in an uncontrolled manner. Unregulated grazing of cattle also leads to soil erosion. There is an ecological imbalance due to loss of balance between land, trees, water and animals. There is less resistance to face famines under the situation of biotic pressures of growing livestock population and limited natural resources of fodder and water in the state.

Political Reasons

The government has given more attention to temporary relief works rather than to the creation of productive community assets on a long-term and sustainable basis. The utility of relief work is of short-term nature. Had we fought with famines in a planned manner from the very beginning, it could be possible to send back this "unwanted guest" to its home. But, due to administrative and other policy failures, it just refuses to go back and continues to bother the desert regions every now and then.

Government Measures to Solve the Problem

The problem of famine is a long-term problem, therefore, long-term planning would be necessary to solve this problem, but when famine occurs, we have to devise short-term measures to tide over the difficult periods. Therefore, the problem, by its nature, needs to be tackled both on short-term as well as long-term bases.

Short-term Strategy for Drought Management

Government starts famine relief works to face the situation as and when it occurs. The state government demands financial assistance from the Centre. On the basis of financial resources, various types of public works are started such as construction of roads, schools, wells, tanks, their maintenance and repair works, soil conservation etc. All possible efforts are made to increase the availability of drinking water to people and livestock. The supply of fodder is arranged from distant sources. Thus, an effort is made to save the people and livestock from all sorts of difficulties. In this process of fighting famines and droughts, opportunities for employment, income and construction of durable community assets are created to a great extent.

The state government runs special programmes for reducing the intensity of famines, which are as follows:

(a) Drought-Prone Area Programme (DPAP): This is a Centre-sponsored scheme financed by the Centre and state on a 50:50 basis. The objective of the programme is to ensure optimum utilisation of land and water in drought prone areas and also to minimise the ill-effects of drought and scarcity conditions on the economy of these areas. Presently, DPAP is being implemented in 32 blocks of 10 districts. This programme has been implemented

on a watershed basis. Each watershed will cover an area of 500 hectares in every village in a phased manner. Programmes of soil conservation, irrigation, afforestation and grassland development have been undertaken under it.

(b) Desert Development Programme (DDP): This programme is totally financed by the Centre. Its objectives are to arrest the process of desertification and to use the limited resources for drought proofing as well as for improving the economy of the desert areas. The programme covers 85 blocks of 16 desert districts. This programme includes activities related with desert tree plantation, groundwater, fodder development, livestock development, water supply and rural electrification etc.

Long-term Strategy for Drought Management

(a) Extension of irrigation facilities: Famines can be contained effectively by increasing irrigation facilities over a wider area. There should be a detailed long-term water management policy for the state, as well as for the whole nation, including ground and surface water and water requirements for different purposes. Indira Gandhi Nahar Project (IGNP) has become a lifeline in eastern desert of Rajasthan. Joining of rivers of the nation could be a boon for Rajasthan, as well as for flood-affected areas of the nation, as that excess flood water could be used in Rajasthan for irrigation, after joining the rivers.

(b) Drainage facilities should be improved: Due to lack of drainage facilities, the problem of waterlogging and other related problems are coming up in the IGNP area. Due to lack of drainage facilities, swampy and marshy lands get developed and give rise to problems like increasing salinity in soil, which is harmful for agriculture operations in the state.

(c) Effective coordination between various programmes: There are various schemes and programmes sponsored by the state as well as the Centre. There must be coordination between these various developmental programmes, and these programmes should be tailormade according to the local resources and needs so that the rural population could be benefited at large and to the maximum. Selection of projects must be such as to help directly in reducing the impact and incidence of droughts and famines in the various parts of the state. Funds should be spent judiciously on employment generation, protection of cattle wealth, fodder

development and increasing its availability and developing agriculture in all possible ways.

(d) Stabilisation of sand dunes: Efforts should be made to stabilise sand dunes so that wind erosion is reduced to a great extent. Fodder trees like *Khejra* should be planted. *Ber*, shrubs, Rohida trees, etc., are very useful for desert areas. The production of fodder should be increased by developing crops like *moth* and *guar* seed. Foreign technical and financial assistance should also be utilised and encouraged for the purpose.

(e) Non-farming activities should be promoted: It is not safe to depend on agricultural or farm activities alone, therefore, non-farming activities should be developed in rural areas so that people may get employment in non-farming activities also. The development, telecommunication, etc. would also make the rural economy resilient and capable of facing famines and fighting with them in a more resolute manner.

Conclusion

The aforesaid discussion indicates that the improvement of water management is crucial for drought management in Rajasthan. There should be proper coordination between short-term and long-term measures to solve the problem of droughts and famines in future. The cooperation of voluntary agencies and that of non-governmental organisations (NGOs) would go a long way to mitigate the problem of droughts and famines in the state. Public cooperation is a necessary condition to fight famines and droughts more successfully in the years to come. There should be greater attention towards the creation of durable and productive community assets in planned rural development in future. That alone would reduce the intensity and incidence of famines and droughts in the state.

24

Facing Famine Fearlessly: Services Way!

I.C. SRIVASTAVA

Famine has been a recurring, though infrequent, phenomenon in recorded history. But, the last two hundred years have witnessed both natural and (almost) man-made famine conditions in several countries. Ireland faced disastrous famines in the 1840s resulting in the death of one-eighth of its population. Towards the end of the nineteenth century, India under the British rule struggled through a severe famine and later in the midst of the Second World War in the 1943 Bengal famine thousands of people died despite strenuous efforts made from various quarters. Bangladesh witnessed the worst famine conditions in 1974 while in Ethiopia famine lasted for more than a decade (1973-85). Two socialist countries, viz., China during 1958-61 and Kampuchea in late seventies had their share of famine conditions wherein state guarantees to provide entitlement of work and food failed to provide the means of survival.

The people of Rajasthan, proverbially known as the land of desert, have fought and overcome several periods of famine and drought conditions over centuries. History has it that in the 11th century the region encompassing Rajasthan faced a 'no rain' period for 12 successive years. In the nineteenth century, besides the deadly drought of 1842-43, the year 1899, coinciding with Vikram

Samvat 1956, witnessed the most severe famine, also known as *Chhapania Akal* resulting in death of hundreds of heads of cattle and human beings. Strangely, in the last two decades, only three years (1983-84, 1990-91 and 1994-95) remained famine-free periods. In the year 2001-02, however, almost the entire state population residing in nearly 40,000 villages of 32 districts was in the grip of famine conditions. Although the state government launched a large number of relief works in almost all districts, only a few lakh persons benefited in terms of employment and wages. The state government also provided relief and succour to the destitute and the needy by doling out foodgrains and old-age pension. Fortunately, no human starvation death was reported due to better management as well as cooperation of the people. Human beings survived but large number of cattle deaths could not be prevented.

Though famine and hunger are not modern maladies, it is quite strange and paradoxical that with the sum total of peaking human development and accompanying accomplishments, glories and successes, we have failed to prevent human starvation and deaths despite worthwhile efforts, world-wide, by United Nations and its sister agencies, viz., UNDP, UNICEF and Economic and Social Council. Inadequacies of food distribution and inequalities of opportunities continue to plague the planning processes and administrative abilities of rulers and bureaucrats in a majority of countries. The institution of wage labour in the prevailing market scenario does not guarantee food and other entitlements of food. The state, of course, has an important role to play in combating hunger, poverty, unemployment and undernourishment. Thus, the state can and should evolve policies focused on employment and social security so as to enable all able-bodied persons to work and earn their 'entitlement'.

As defined by the Nobel laureate, Prof. Amartya Sen, "entitlement of a person stands for the set of alternative commodity bundles that can be acquired through the use of the various legal channels of acquirement open to that person." This may include ownership of property and endowment in a private economy besides one's own capacity to work for wages. In an agricultural economy, cultivation of crops, which is its mainstay, becomes what has been termed as 'direct entitlement' for the family based on output and availability of foodgrains. But, a family's purchasing power during famine depends on price of food, which is influenced by production and availability of foodgrains. Management of public distribution system is also a major instrument to combat starvation through governmental action. Paradoxically, adequate availability of foodgrains in the economy is a poor guarantee against starvation deaths. The Bangaladesh famine of 1974 illustrates that paradox vividly in that availability of food that year was higher than in any preceding three-four years. Rural labourers and their families died owing to their inability to earn or command food due to loss of employment and a variety of other reasons besides decline in food production. Prevention of famine, therefore, is determined by the extent of protection of entitlement through promotion of employment generation on a long- term basis.

However, famine relief focusing attention on emergency operations can be distinguished from long-term entitlement protection measures which envisage a plurality of strategies. Massive works programme, launched by the Maharashtra government in 1972-73, successfully prevented severe drought conditions from developing into famine assisted as it was by various international agencies in mounting a feeding programme. The Maharashtra government later continued and converted the works programme into a sustainable, what has come to be called, Employment Guarantee Scheme. Fighting and facing famine as an integral part of long-term and total development strategy, therefore, calls for creation of a feasible, viable and sustainable set of employment opportunities.

In fact, creation of employment and redistribution of income to the poor was advocated and supported by Kautilya in the 4th century BC as part of a sound administrative system to defeat famines. In modern times, the British-appointed Famine Commission recommended two-fold strategy in 1880: (i) framing of region-specific Famine Codes embodying 'authoritative guidelines' to the local administration on the measures needed to anticipate and deal with the threat of famine; and (ii) strategy of entitlement protection based on the combination of guaranteed employment, a subsistence wage and unconditional (gratuitous) relief. Paradoxical though it may appear, our administrative apparatus, which is a British legacy, more often, struggles and muddles through the famine and drought conditions swearing by the same age-old Famine Code guidelines and by launching the so-called famine works mainly of gravel road work or deepening of tanks and water bodies besides construction of some *pucca* structures. In addition, gratuitous relief is provided in terms of foodgrain supplies and cash-dole.

However, a long-term integrated strategy calls forth capacity to envision a comprehensive set of measures for generation of sustainable livelihood opportunities for the poor. In this context, let us recall with conviction, the vision embodied in our President, Dr. A.P.J. Abdul Kalam's book, *India 2020*: "One thing is crystal clear. Indian human resource base is one of the greatest core competencies. It is India's strength. If we can train an unskilled Indian, if we can impart better skills to a skilled Indian and if we can create a more challenging environment for the educated as well as build avenues for economic activity in agriculture, industry and the service sectors, these Indians will not only meet the targets but excel them. Indians not only have a great learning capacity, but most of them also have an entrepreneurial and competitive spirit. Today, there are not enough avenues to channelise this spirit constructively and productively."

In this context, let us now recall and consider the present status of agriculture and avenues of employment in other sectors. Indian population is projected at 1.3 billion by the year 2020. Economic growth and changes in lifestyle lead to higher consumption resulting from higher income. A demand of 340 million tonnes of foodgrains is, therefore, projected by 2020 as a compared to present over 200 million tonnes. As it is, rainfed areas accounting for 70 per cent of the net cultivated areas of the country have not benefited largely from modern development in agriculture. Biotechnology can, therefore, help improve seeds, plants and soil treatment etc. Besides, conservation of water, coupled with improvements in agricultural implements, machinery, water technologies, biofertilisers, etc., are within our country's reach. Remote sensing from space is another helpful instrument for assessing national resources. Finally, post-harvest technologies and agro-food processing activities can help generate numerous employment opportunities and also eliminate storage and transportation losses.

Besides, agri-business activities are opening up a number of services in agriculture sector itself. The strength of Indian economy is thus growing in the vast expanding domestic services market. The features of our growing and healthy economy can be summarised as follows : (i) fast expanding domestic market, (ii) growth of wage opportunities, (iii) growing scope of self-employment, (iv) expanding informal economy, (v) fast expanding service sector, (vi) emerging non-cropping economic activities in rural area, (vii) increase in small and medium entrepreneurship, and (viii) financial services revolution. Against this backdrop, the real issue is to draw out the best potential of our people, especially in the rural areas, by motivating them and by training them in various skills. As the distinguished economist, Amartya Sen aptly puts it: “The central issue is to expand the social opportunities open to people… But the creation of social opportunities on a broad basis requires much more than the freeing of markets. It calls, in particular, for expansion of educational facilities and health care for all and public provisions for nutritional support.”

As we all know, in India, 60-70 per cent of the workforce live in rural areas and depend on agriculture directly or indirectly. But slow economic growth does not allow large absorption of the rural poor in other economic activities. Hence, a large number of people are forced to survive as marginal farmers or marginal landless labourers. Besides, a growing number of unskilled educated youth are unemployed and, are waiting in the wings as it were, in the rural areas to come forward and play an active role in our expanding economy. The rural population has four distinct segments based on access to land: (i) large land owners, (ii) salaried households, (iii) agricultural and non-agricultural wage labourers, and (iv) marginal farmers. As it is, rural economy reflects vast and deep poverty which manifests itself in low consumption, malnutrition, illiteracy, insecurity, low expectancy of life, powerlessness and low self-esteem.

Considering the size of rural credit and outreach of banking system in the context of employment generation, a study conducted by Ramola and Mahajan (1995) revealed that formal bank credit accounted for only 16 per cent of the total credit whereas 84 per cent of credit needs were met by informal credit system comprising traders, shopkeepers, moneylenders, landlords, friends and relatives. But, now, over 7.8 million poor households have access to banking services through 4,61,478 Self-Help Groups (SHGs) all over the country. Today, over 2,000 NGOs and 17,000 branches of 444 banks are associated with SHG-bank-linkage programme of NABARD. Thus, there are immense possibilities of generating employment for rural women through this mechanism of SHGs, already in place as a huge sum of over Rs. 1,026 crores has already been disbursed to members of SHGs, empowering rural women with new found self-confidence to take change of their lives and livelihood.

The success of SHGs in rural areas reveals that new winds of change and development have started blowing through the villages and rural workforce is now clearly moving towards nonagricultural activities. While the share of agriculture in GDP has declined from 44.8 per cent to 27.6 per cent, the share of employment in agriculture has sharply come down to 60.2 per cent from 73.9 per cent in the country as a whole. However, in Rajasthan, total workers in agriculture, including cultivators and agricultural labourers, constitute nearly 66 per cent of the workforce (cultivators 55.37 per cent and agricultural labourers 10.63 per cent). Hence, the need to wean away rural citizenry from cultivation to agri-business and services sector including processing, grading and packaging, general marketing and marketing for export cannot be overemphasised as the future of the country lies in the growing services sector which has opened up huge possibilities for a variety of service activities. It is a good augury that central as well as state governments have set up institutes of cooperative management to give a boost to agri-business entrepreneurship.

However, it is now widely accepted that a radically different and revolutionary approach to catapult the growth of the farm sector is an urgent imperative to bring about rural transformation. It is also true that government alone cannot shoulder the entire responsibility. Nonetheless, it cannot slacken its efforts. The best way out is to involve and rope in all stakeholders. In this context, it is worthwhile to mention the initiative taken by ITC launch IT-driven "e-choupal" project—the largest internet-based real-time information and customised knowledge to improve farmer's decision-making ability in order to align farm output to market demand and improve productivity. The project, through the use of IT, aggregates the farmers' demand for inputs and services by creating, in effect, a virtual producers' cooperative, which farmers can access easily for obtaining the best quality farm inputs and services. The project is now moving towards setting up a direct marketing channel with efficient price discovery and low transaction costs in output marketing. Thus, the multidimensional use of information technology and internet is on way to empowering farmers with real-time linkage to world market. The project provides an outreach to 6.50 lakh farmers in the 6,500 project villages through 1,020 'e-chaupals' in four states and gives real-time information on local weather and knowledge about farm practices in addition to 'virtual mandi' system and freedom to take decisions. ITC plans to cover one lakh villages in 15 states in next six to seven years and cater to wider range of crops including cotton, spices, and some others in horticulture.

ITC's 'e-choupal' project also illustrates how e-governance can help improve interface of administration with the general public and bring in transparency and accountability in the functioning .of officialdom at various levels. To illustrate, 'Janmitra' pilot project, now functional in Jhalawar district, has set up one-stop contact point (computer kiosk) for citizens to access various government services and informations including online and application forms and issue of copy of land records and mutation (*jamabandi* and *namantarkaran*), grievance redressal and information about development works, public distribution system, health services, government schemes and agricultural and *mandi* rates.

A similar but more inclusive project, launched by Aks Broadband Co. and named 'Gramdoot' presently covering nearly 100 villages, has been moderately successful in Jaipur district. Apart from single-window e-governance services as provided by 'Janmitra', Gramdoot also renders convergent services like cable, internet web-conferencing and online computer education besides being equipped with photoshop for digital photos. The project has run into rough weather due to paucity of funds, unwillingness of banks to finance and government's inability to provide necessary guarantee.

Future beckons us towards IT-driven services, which will necessarily open new vistas for large employment opportunities. Technology savvy, President Kalam has, however, rightly raised the question: Can a country survive on services? He himself has answered it thus: "This may be true for smaller countries. A country like India cannot hope to build its future on the services sector alone though it can be and will be a major component of the economy. India cannot afford not to build its strengths in agriculture for reasons of food and nutritional security nor can it afford to ignore manufacturing strengths for reasons of economic and national security. Based on the strengths of these two sectors, it can build a major economic infrastructure for the services sector and use it to generate great wealth and employment for her people." Dr. Kalam, then, goes on to enumerate the potential and upcoming services as follows: (a) financial services, (b) marketing communication services, (c) marketing logistics trading and distribution, (d) trade promotion services, (e) human resource development, (f) government administration, (g) security services, (h) repair and maintenance, (i) tourism and hotel, (j) leisure and sports, (k) old-age care services, and (1) preventive health care services.

Thus, in the emerging scenario, training in computer application for rural development, including agriculture and marketing of agricultural produce, is crucial for IT-driven services in rural areas. Already, Aptech and NIIT training outfits have opened branches and franchises in a large number of urban areas. In Rajasthan, Computer Consultancy and Accounting Analysis Ltd. (CAL) is rendering yeoman service in training young men and women with its branches in almost all the district headquarters besides a few smaller towns. It has already trained over 3,000 young men and women in computer applications, and most of them are now awaiting employment opportunities.

While futuristic information technology will be the prime driving force to give necessary push to multiple services, rural people will have to rise to the occasion to take advantage of various technology packages and help themselves by exploring new avenues of work and channels of interdependence. For instance, there is vast scope of work in handling of rural litigation and government information services by the educated unemployed who can also be trained in writing various applications and fighting for local problems and causes besides grievances redressal matters likely to come up before them for pursuing with government functionaries at various levels. Computer kiosks on the lines of STD/PCO booths can be acquired by the young men and women with bank loans and subsidy given by government under Prime Minister's Rojgar Yojana and DPIP in the areas covered by AKS Broadband Company.

It is now widely acclaimed and accepted that information technology can be used as engine of growth, i.e., the driving force for providing impetus to various services both in rural and urban areas. But, a set of sequential and simultaneous measures need to be taken. For instance, in order to improve labour absorption in agriculture, year-round cultivation along with agri-based value-added products will open alternative investment opportunities for the employers and labour opportunities for the agricultural labourers leading to reallocation of resources in the rural areas in order to enable them to withstand the rigours of famine and scarcity in future.

Simultaneously, launching of rural enterprise development projects will necessitate a different strategy. For instance, entrepreneurship-led strategy will need to focus on training in decision-making to small entrepreneurs. On the other hand, productivity and market-led strategies call for emphasis on competitiveness and output quality together with identifying viable products conforming to market needs and opportunities which need exploring and reinforcing by developing sourcing and market relationship. Finally, resources-led strategy would imply beforehand analyses of new materials which can be uniquely and profitably exploited by rural enterprises. These issues, together with issues of linkages and coordination, deserve to be discussed threadbare in local and regional workshops and interactions before strategies for operationalisation can be evolved in specific regions based on felt needs.

Against the backdrop of vast and immense possibilities of opening up services sector, we observe that multinationals backed by the developed western countries are eyeing our domestic rural markets offering millions of consumers financial services and consumer goods. Therefore, it is high time that we ourselves develop market research and communication strategy to strengthen indigenous production of goods and services. Agriculture extension services, which ushered in 'Green Revolution' in sixties and seventies leading to self-dependence in foodgrain production, should be strengthened to bring about and sustain Second or Continuing Green Revolution, as Dr. M.S. Swaminathan calls it, which would comprise the agri-package of practices to be transferred from lab to land through information technology to be placed at the disposal of semi-skilled common men and women as is already being done with great success in Tamilnadu by Dr. M.S. Swaminathan Foundation.

Investment in rural areas and development of rural entrepreneurship will receive a boost only if a network and convergence of training by government agencies and non-government organisations can be evolved and integrated into a strategy which would seek to obliterate the dichotomy of functioning styles of government departments/agencies as opposed to NGOs.

The region- and area-specific action plan, comprising capacity building and training projects of all role players and stakeholders on the one hand and skill development for the unskilled and upgrading of skills for the semi-skilled in training institutions on the other, will have to be prepared and financed by the banking and other lead financing institutions under the aegis of district and local administration. The action plan and strategy will be based on intense and wide consultations with local groups consisting of representatives of PRIs, NGOs and line department in workshops to be held in each development block or a group of blocks in districts with potential market and resources.

Much will depend, nonetheless, on the initiative taken by administrators of the area and the political will of the local leaders and representatives of PRIs to build up services sector activities as guarantee against onslaught of famine in future.

As services sector holds promise of bright future in terms of fast expansion of our economy, the sooner we initiate and take convergent set of measures as outlined above, the better it would be for the multitudes of able-bodied youth with vibrant energy and eager minds to play an active role in the economy. To sum up, we may recall the clarion call given by our President Dr. Kalam, in his book, *India 2020*: "As a country of one billion, we should gear ourselves to take up the opportunities offered by services sector in our march towards an India where every Indian will have wealth and well-being."

List of Contributors

A.N. Lahiri, formerly associated with Central Arid Zone Research Institute, Jodhpur.

Anirudh Krishna, Assistant Professor of Public Policy Studies, Terry Sanford Institute of Public Policy, Duke University. Before turning to academia, he worked for 13 years with the Indian Administrative Services.

Bhupendra Hooja, held many senior positions in the Government of Rajasthan as an IAS officer. He continued writing after retirement and was Chief Editor, *Indian Book Chronicle.*

C.M. Mathur, IFS, retired as Chief Conservator of Forests, Government of Rajasthan, Jaipur.

Harsh Sethi, Consultant, Business India Group and Consultant Editor, *Seminar.*

H.S. Mann, former Director, Central Arid Zone Research Institute, Jodhpur.

I.C. Srivastava, formerly Chairman, Board of Revenue, Ajmer, served in the IAS in Rajasthan.

J.P. Kanawa, Research Officer, HCM Rajasthan State Institute of Public Administration, Jaipur.

K.A. Shankarnarayan, former Director, Central Arid Zone Research Institute, Jodhpur.

Kamala Jain, Associate Professor of History, Mohan Lal Sukhadia University, Udaipur.

Kanwar Sain, former Chairman, Central Water and Power Commission, New Delhi.

L.D. Ahuja, formerly associated with Central Arid Zone Research Institute, Jodhpur.

M.K. Khanna, IAS, has served in Government of Rajasthan and Government of India, and is presently Principal Advisor, Planning Commission of India, New Delhi.

Nathu Ram Mirdha, former Chairman, National Commission of Agriculture, Government of India.

R.P. Singh, formerly associated with Central Arid Zone Research Institute, Jodhpur.

Rakesh Hooja, IAS, is Additional Chief Secretary to Government of Rajasthan and Director, HCM Rajasthan State Institute of Public Administration, Jaipur.

Rakshat Hooja, Associate Director, Minnesota Studies in International Development, India Program, Jaipur

Ramesh K. Arora, Visiting Professor, HCM Rajasthan State Institute of Public Administration, Jaipur, has worked as Professor and Dean, Social Sciences, University of Rajasthan.

S.P. Malhotra, formerly associated with Central Arid Zone Research Institute, Jodhpur.

S.P. Vishnoi is a retired IAS officer of Rajasthan cadre and a former Director, HCM Rajasthan State Institute of Public Administration, Jaipur.

Sanjay Sharma, formerly UGC Senior Research Fellow in the Department of Sociology, University of Rajasthan.